An Intro...
Basic Astron...
(with Space Photos)

Chris McMullen, Ph.D.
Northwestern State University of Louisiana

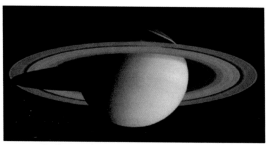
An Introduction to Basic Astronomy Concepts (with Space Photos): A Visual Tour of Our Solar System and Beyond.

Copyright © 2012 Chris McMullen

Astro Nutz.

ISBN-10: 1478169389.
ISBN-13: 978-1478169383.

Chris McMullen

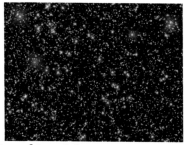

Northwestern State University of Louisiana

Contents

1 Overview of the Solar System

1.1 The Solar System

Our solar system consists of a star, which we call the sun, eight planets which orbit the sun, over a hundred moons which orbit the planets, and numerous dwarf planets, asteroids, and comets which orbit the sun. The orbits of the planets, moons, dwarf planets, and asteroids lie in nearly the same plane, such that our solar system has the approximate shape of a plane. Most of the solar system is actually empty space – there are very large gaps between the astronomical bodies.

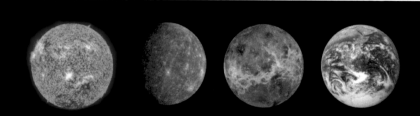

Our Solar System

The order of the planets from nearest to furthest from the sun, illustrated in the previous figure, is Mercury, Venus, Earth, Mars, Jupiter, Saturn, Uranus, and Neptune. Almost all of the asteroids lie in the Asteroid Belt between Mars and Jupiter. Many comets reside in a similar belt beyond Neptune, called the Kuiper (Kie-per) Belt. (Note that the "Kui" in "Kuiper" is pronounced with a long "i," as in "pie.") The dwarf planet Pluto resides in the Kuiper Belt. Due to Pluto's small size, eccentric orbit, and its location in the Kuiper Belt, it is no longer considered by astronomers to be a planet like Neptune, but is now considered to be one of many dwarf planets that reside in our solar system. Other large bodies in the Asteroid Belt and Kuiper Belt are also considered to be dwarf planets.

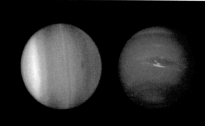

The Asteroid Belt divides the planets into two groups: The terrestrial planets – Mercury, Venus, Earth, and Mars – lie between the sun and the Asteroid Belt, while the Jovian planets – Jupiter, Saturn, Uranus, and Neptune – lie between the Asteroid Belt and the Kuiper Belt.

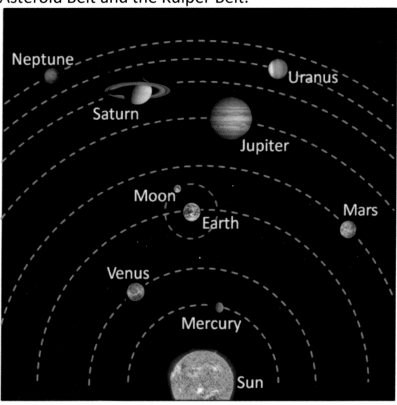

The terrestrial planets are closer to the sun, have higher surface temperatures, are small planets, have only a few moons all together (Mercury and Venus have no moons, Earth has one, and Mars has two very small moons), and are very rocky. The Jovian planets (also called gas giants) are further from the sun, are very large planets, have many moons (over 100 all together), and are gaseous (made of hydrogen and helium).

Every planet makes two types of circles. (1) One type of circle is a revolution. The orbit of a planet around the sun is called a revolution. When we say that a planet revolves around the sun, we mean that it orbits the sun. (2) Each planet also rotates on its axis while it revolves around the sun. The rotation of the planet refers to its spin. When we say that a planet rotates, we mean that it spins on an axis.

It's very important not to confuse the terms rotate and revolve in astronomy. For example, Earth completes a rotation about its axis every 24 hours, while it takes 365 days to complete a revolution about the sun. The Earth rotates 365 times in a single revolution.

The planets orbit the sun in elliptical orbits. The sun lies at a focus, which is not far from the center of the ellipse (it's actually in the center of mass of the system, but since the sun is by far the most massive object in the solar system, the focus and center of mass are very close to one another).

The Earth's closest approach to the sun is called perihelion (since the prefix "peri" means near) and the furthest point is called aphelion ("ap" and "ab" mean away). The moon's closest approach to the Earth is called perigee ("helion" refers to the sun, while "gee" and "geo" refer to the Earth) and its furthest point is called apogee.

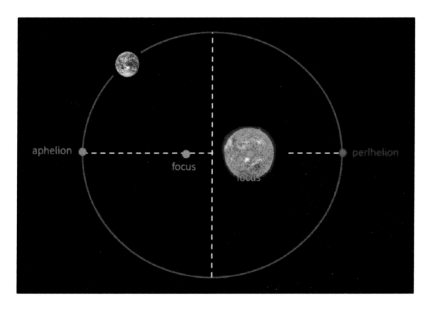

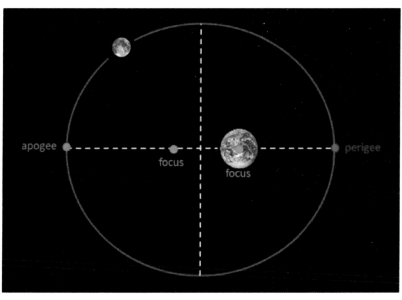

The sun is by far the largest object in the solar system. The sun is 108 times larger than the Earth, and is about 10 times larger than Jupiter.

Jupiter is the largest planet in the solar system. It is 11 times larger than the Earth. The second largest planet is Saturn, which is about 9 times larger than Earth. The smallest planet is Mercury (now that Pluto is considered to be a dwarf planet rather than a planet).

The figure on the following page shows the sizes of the sun and the planets to scale (but the distances between them is not shown to scale).

1.2 The Sun

The sun is the largest and hottest object in our solar system. It is a star, which radiates an enormous amount of light each second. The Earth receives just a tiny fraction of the sun's energy, yet it is plenty of energy to keep the Earth warm and to sustain life. The sun is our primary energy source, and life on Earth depends on the sun's energy supply. The surface temperature of the sun is 5800 Kelvin. That's more than a high enough temperature to melt any metal.

The sun is 108 times larger than the Earth in size, yet you could fit 1,000,000 (one million) Earths inside of the sun! How is this possible? The volume of a sphere is proportional to the cube of the radius:

$$V = \frac{4}{3}\pi R^3$$

This means that the factor of 108 (from the radius) gets raised to the power of 3 (we say that it is cubed) when calculating the volume. 100 cubed = 100 times 100 times 100 = 1,000,000. Therefore, the sun has a million times as much volume as the Earth, even though it only has about 100 times the radius of Earth.

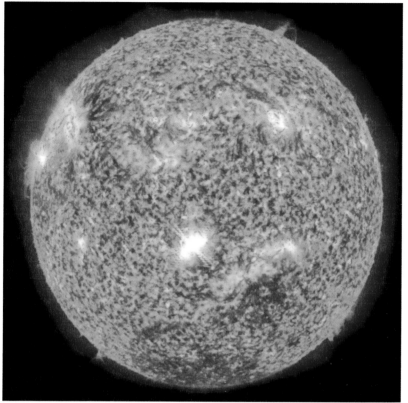

The sun is also incredibly massive. The sun has 333,000 times as much mass as the Earth. Amazingly, 99.8% of the solar system's total mass lies in the sun!

The sun is less dense than the Earth. On average, the sun is 1.4 times as dense as water at Earth's average surface temperature and pressure. Compare the sun's average density to that of the Earth, which has an average density that is 5.5 times that of water. The sun is composed primarily of hydrogen (H) and helium (He), whereas the Earth is very rocky.

Density is defined as mass divided by volume. Lead is a very dense material – it sinks when placed in water. Wood is not very dense – it floats on water. Helium has very little density – it even floats in air (that's why a helium balloon rises upward).

Hydrogen and helium usually have very little density. Near the surface of the Earth, their density is so small that they rise upward – they are less dense than air. However, the sun is so massive that it has a very strong gravitational pull, which causes hydrogen and helium to be much more dense in the sun than they are on Earth. This is why the sun's average density, 1.4 grams per cubic centimeter (g/cc), is more dense than water (1 g/cc), even though on the Earth hydrogen and helium are much less dense than air.

The sun is so hot that hydrogen and helium exist in the form of ions – atoms that have gained or lost electrons through collisions, which therefore have a net electric charge. The sun is not considered to be a gas (or liquid or solid), but is plasma. A gas consists of neutral particles, whereas plasma consists of ions. At such extreme temperatures, very frequent, energetic collisions cause atoms to gain or lose electrons and exist as ions rather than neutral atoms.

Every second, the sun converts four million tons of mass into energy by converting hydrogen nuclei into helium nuclei. Specifically, four hydrogen

nuclei bind together to form a helium nucleus in an exothermic reaction (a reaction that emits energy). Such a process is called nuclear fusion. This nuclear energy is the energy that causes the sun to shine. The difference in mass between four hydrogen nuclei and one helium nucleus causes energy to be released in nuclear fusion through Einstein's famous equation, $E = mc^2$, which expresses an equivalence between mass and energy.

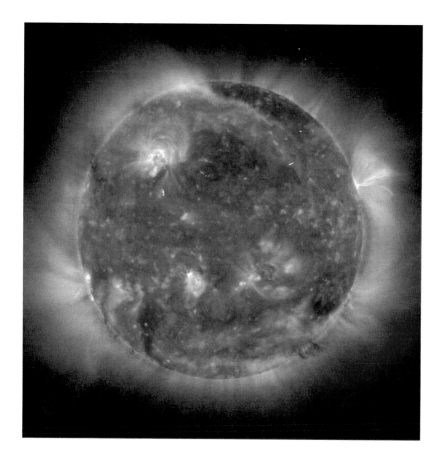

Nuclear fusion occurs in the sun's core, where the temperature is about 15 million Kelvin. That's the hottest part of the sun. Energy released from the core makes its way to the surface by passing through a radiation zone and convection zone. The visible surface of the sun that we see is called the photosphere, and has a much cooler temperature of about 5800 Kelvin. Sunspots that can be seen on its surface are cooler regions with a temperature of about 4000 Kelvin; these are areas where there are strong magnetic fields. The number of sunspots that appear on the sun fluctuate in 11-year cycles. Beyond the photosphere are two layers of atmosphere – the chromosphere and corona. The corona is the outer layer of the sun's atmosphere. Beyond the corona, the solar wind takes charged particles (ions) away from the sun. Charged particles from the solar wind that reach Earth produce colored lights in the northern and southern hemispheres of Earth called the Aurora Borealis and Aurora Australis, respectively (also called the Northern and Southern Lights). Sudden bursts of these charged particles, linked to the sunspot cycle, can also cause disruptions in electric power and satellite communication on Earth.

The ultraviolet (UV) photos of the sun shown here show the solar activity that is constantly occurring near its surface. Don't look directly at the sun because direct sunlight has enough intensity to

permanently damage your eye in very little time; it can burn out your retinas in eight minutes, and do very serious damage in much less time.

1.3 Mercury

Mercury is the closest planet to the sun. Now that Pluto is considered to be a dwarf planet instead of a planet, Mercury is also the smallest planet. Its radius is 0.4 times that of Earth's, and its orbital radius is also 0.4 times that of Earth's. Mercury's mass is 0.055 times Earth's mass. Mercury is a rocky planet, with a density (5.4 g/cc) almost as great as Earth's (5.5 g/cc).

Mercury has virtually no atmosphere. It would be possible to stand on Mercury during daytime and see stars by looking away from the sun (whereas on Earth you only see stars during nighttime because of its thick atmosphere). Mercury is heavily cratered, like the moon, because it has a very thin atmosphere and is geologically inactive.

Mercury also has the most extreme day/night temperature variations. It is 700 Kelvin during the day (that's much hotter than boiling water), but only 100 Kelvin at night (that's about 200 K colder than the freezing point of water).

One 'year' on Mercury lasts 88 Earth days – that's how long it takes to orbit the sun. One 'day' on Mercury lasts 59 Earth days – that's how long it takes to rotate about its axis. Mercury spins very slowly: It barely completes one rotation about its

axis before completing its orbit around the sun. Every day and night on Mercury lasts about one-third of a Mercury year. The slow rotation (but the revolution is fast) is due to strong tidal forces from the sun.

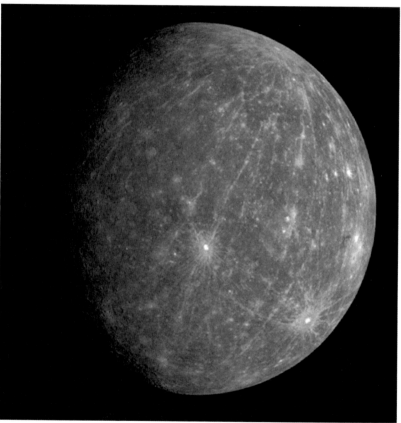

Mercury's axis of rotation is not tilted like Earth's, which means there wouldn't be seasonal variations between the northern and southern hemispheres. Like the moon, Mercury is desolate.

Mercury does not have any moons or rings. It is composed mostly of rocks and metals.

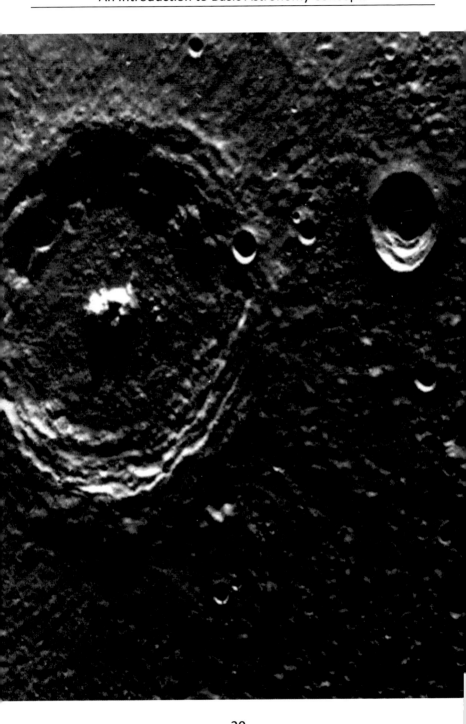

1.4 Venus

Venus is the second planet from the sun. It is almost as large as Earth, with a radius that is 0.95 times that of Earth's, while its orbital radius is 0.7 times that of Earth's. Venus's mass is 0.82 times Earth's mass. Venus is also a rocky planet, with a density (5.2 g/cc) almost as great as Earth's (5.5 g/cc).

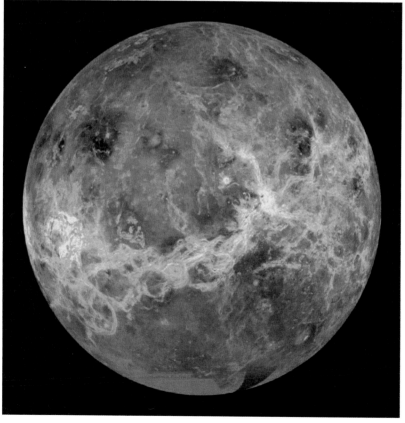

Venus has the hottest average surface temperature – 740 K. The surface is so hot that the spacecraft that have landed there to take pictures were only able to take and send photos for a short period before the instruments became too hot to function. Venus's average surface temperature is actually a little hotter than Mercury's average daytime temperature, even though Mercury is closer to the sun. The reason for this is that Venus has an extreme greenhouse effect (trapped solar radiation that warms a planet's surface and cools its atmosphere). This illustrates the extreme potential of global warming (even worse for Venus, which is closer to the sun than Earth).

Venus is always covered with clouds. The pictures that we have of Venus either show its clouds or use radar to produce an image of the surface beneath the clouds, except for close-up images of the surface taken by landing spacecraft. The previous photo was simulated from various data. The following photo, which shows the clouds of Venus with a UV filter, was taken by the Mariner 10 in 1974.

Venus is very similar to the Earth in size, composition, atmospheric phenomena (except for being much more extreme on Venus), and geological activity (except that it lacks plate tectonics – which drive Earth's continents). However, Venus is extremely hot and lacks liquid water.

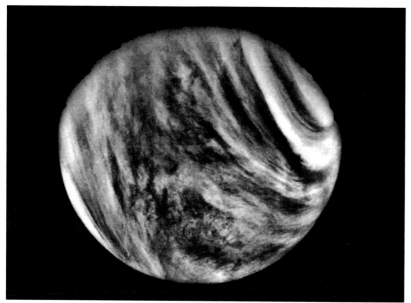

One 'year' on Venus lasts 225 Earth days. One 'day' on Venus lasts 243 Earth days. It is the only planet where a 'day' is longer than a 'year' – i.e. Venus spins so slowly that it actually travels all the way around the sun before spinning one time about its axis. Another peculiarity is that Venus is the only planet that rotates backwards on its axis. Its revolution around the sun is in the same direction as the other planets, but the way that it spins on its axis is opposite to the way that the other planets rotate. This unusual rotation could be explained by a possible collision in Venus's history.

Venus's axis is tilted 177°. Like Mercury, Venus does not have any moons or rings and it is composed mostly of rocks and metals.

1.5 Earth

E arth is the third planet from the sun. Earth's radius is 6.4×10^6 m (or 6400 kilometers), its average orbital radius is 1.5×10^{11} m (or 150 million kilometers), and its mass is 6.0×10^{24} kg (or 6 trillion trillion kilograms). Earth's average orbital radius also equals 1 astronomical unit (AU). Since solar system distances are very large, it is convenient to measure them in AU rather than km. It takes about 8 minutes for light to travel from the sun to the Earth at a rate of 300,000,000 (three hundred million) m/s. Earth, also a rocky planet, is the most dense (5.5 g/cc) planet in our solar system.

Earth's average surface temperature is 290 K, which equates to 17°C or 63°F. This is just the average surface temperature. Of course, it gets much colder near the poles and warmer near the equator. Earth also experiences significant seasonal effects.

One year on Earth is 365 days, and each day is 24 hours. Earth's rotation is much faster than that of Mercury and Venus, while its revolution is somewhat slower.

Earth's axis is tilted 23.5°. This causes significant seasonal effects. The tilt of Earth's axis means that the northern hemisphere receives more

direct sunlight during the summer and less during the winter, and also explains why it is summer in the southern hemisphere when it is winter in the northern hemisphere, and vice-versa. We will discuss the cause of the seasons in more detail in Chapter 4.

Earth is one of only two terrestrial planets to have a moon, and is the only terrestrial planet to have a very large moon. (Mars has two smaller moons. Also, note that some of the Jovian planets' moons are larger than Earth's moon.) Earth's moon

is believed to have been captured during a collision early in its history. As with all of the terrestrial planets, Earth does not have any rings.

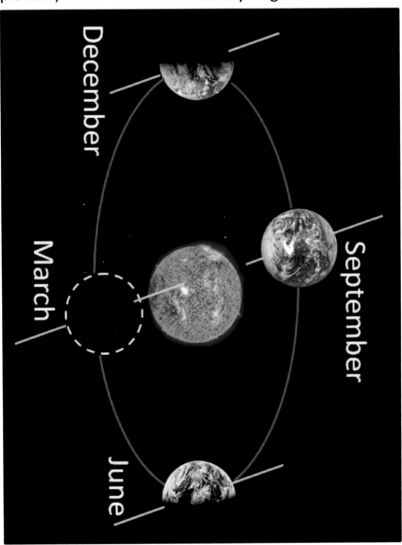

The moon has very little atmosphere and is geologically inactive. Its surface is heavily cratered. However, there are large regions that are smooth,

which appear dark in the previous photo. We call these maria, which means seas. (Mare is singular, maria is plural.)

In 1969, Neil Armstrong became the first person to walk on the moon. Neil Armstrong and Buzz Aldrin walked on the moon that day as part of NASA's Apollo 11 mission.

Astronauts drove NASA's lunar rover on the surface of the moon during the Apollo 15, 16, and 17 missions in 1971-1972. (Contrary to the recent movie, NASA did not launch an Apollo 18 mission.)

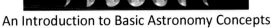

Water covers 71% of Earth's surface. Earth is geologically active, including volcanoes and plate tectonics, and also has very active weather, including hurricanes. The seven continents – North and South America, Europe, Asia, Africa, Australia, and Antarctica – are believed to have originally been part of one original supercontinent – called the Pangaea – which have slowly drifted apart over several million of years.

1.6 Mars

Mars is the fourth planet from the sun. It is smaller than Earth, but larger than Mercury, with a radius that is 0.53 times that of Earth's, while its orbital radius is 1.5 times that of Earth's. The separation between Earth's and Mars's orbits is the largest of any consecutive terrestrial planets (but there are greater separations among the Jovian planets). Mars's mass is about one tenth that of Earth's (more precisely, it is 0.11 times Earth's mass). Mars has the smallest density (3.9 g/cc) of the terrestrial planets (but is much more dense than the Jovian planets).

The average surface temperature of Mars is 220 Kelvin, so the surface temperature is below freezing, on average. Mars has an axis tilt of 25.2°, which is slightly more than Earth's (23.5°). In addition, Mars has a more noticeably elliptical orbit than Earth (which is much more circular). This combination of significant axis tilt and a more elliptical orbit produces interesting seasonal effects. Mars travels faster when it is closer to the sun and slower when it is further from the sun according to Kepler's second law (described in Section 7.2). The result is that the northern hemisphere experiences a long, cool summer and a short, mild winter, while

the southern hemisphere experiences a short, warm summer and a long, cold winter.

One 'year' on Mars lasts 1.9 Earth years. One 'day' on Mars is slightly longer than an Earth day – a 'day' on Mars lasts 24.6 hours.

Mars has two moons, named Phobos (Foebus) and Deimos (Die-mus), which are very small. Phobos, the larger of the two moons, is shown in the following photo. Phobos measures 11 km across, which is about the size of a city. It is probable that these moons were captured from the nearby

Asteroid Belt. Like the other terrestrial planets, Mars does not have any rings and it is composed mostly of rocks and metals.

The largest volcano in our solar system, Olympus Mons, can be found on Mars. Olympus Mons measures 600 kilometers across (that's about the same area as Colorado) and is 26 kilometers high (compare to Mount Everest, which is 8.8 kilometers tall). A top view of Olympus Mons is shown in the following photo.

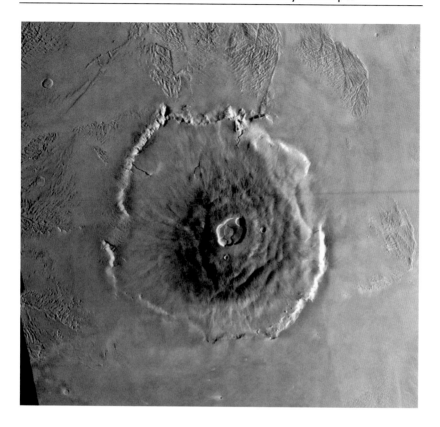

There is also a grand canyon on Mars, called Valles Marineris. It is about four times deeper (up to 7 kilometers) than Earth's Grand Canyon (1.6 kilometers deep), and is 4000 kilometers long – that's about the distance from Seattle, Washington to Portland, Maine across the continental United States. You can see Valles Marineris running horizontally across the center of the following photo. Valles Marineris provides evidence that Mars has plate tectonics (though not on the same global scale as Earth).

The polar ice caps on Mars contain both dry ice (frozen carbon dioxide) and ice (frozen water). Examination of the geographic features of Mars – such as dried riverbeds – show evidence that Mars had liquid water in its distant past.

1.7 Jupiter

Jupiter is the fifth planet from the sun. It is the largest planet in our solar system, with a radius that is 11 times that of Earth's (but about 10 times smaller than the sun's radius), while its orbital radius is 5.2 times that of Earth's. Jupiter has 318 times as much mass as Earth (yet about 1000 times less mass than the sun). Jupiter, like all of the Jovian planets, has a density (1.3 g/cc) that is small compared to the terrestrial planets.

The average surface temperature of Jupiter is 125 Kelvin (compare to 273 Kelvin, which is the freezing point of water at standard pressure). Like all of the Jovian planets, Jupiter does not have a solid surface, since it is composed of hydrogen and helium gas. Hence, we say that the Jovian planets are gas giants. Jupiter has very little axis tilt (3.1°).

One 'year' on Jupiter lasts about 10 Earth years. Jupiter spins faster on its axis than any other planet – one 'day' lasts just 9.9 hours. Jupiter has a slow revolution, but fast rotation (contrast with Mercury and Venus, which have fast revolutions, but very slow rotations). The largest planet in our solar system, Jupiter, also spins the fastest on its axis (but does not revolve the fastest).

The most famous feature of Jupiter is the Great Red Spot. The Great Red Spot, visible in the photo above and enlarged in the following photo, is a very large hurricane (about twice as large as Earth) in Jupiter's atmosphere, which has existed for at least a couple of hundred years. On Earth, hurricanes eventually come across land, where they lose their energy. However, Jupiter has no land, so hurricanes may be very long-lived.

Jupiter has (at least) 63 moons. The four largest moons of Jupiter are Ganymede (Gan-i-mead), Callisto (Kuh-lis-toe), Io (Eye-oh), and Europa (You-row-puh). Ganymede is the largest moon in our solar system. Io's surface has active volcanoes all over it. Io is more volcanically active than any other object in our solar system because Io is heated from strong tidal forces exerted by Jupiter. Europa's surface is icy. In the case of Europa, tidal stress has produced numerous cracks on its surface and tidal heating may have melted the ice enough to produce a subsurface ocean. Callisto is a giant ball of ice, with craters instead of cracks because it does not

experience tidal heating (it does not participate in the orbital resonances of the other three large moons).

Jupiter, like all of the gas giants, has rings, but they are very faint compared to Saturn's prominent ring system. In the following photo, the rings appear as a pair of orange lines, while the curved colors mark the edge of Jupiter. This photo was taken during Jupiter's shadow using color filters.

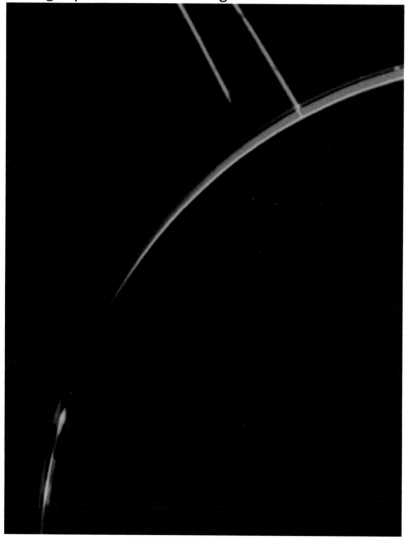

1.8 Saturn

Saturn is the sixth planet from the sun. Saturn is the second largest planet in the solar system, with a radius that is 9.4 times that of Earth's, while its orbital radius is 9.5 times that of Earth's. Saturn has 98 times as much mass as Earth. Saturn has the lowest density of any planet (0.7 g/cc) – it's even less dense than water at standard temperature and pressure (1 g/cc).

The average surface temperature of Saturn is 95 Kelvin, which is well below freezing. Like the other Jovian planets, Saturn is composed of hydrogen and helium gas. Saturn has an axis tilt of 26.7°, which is a little more than Earth's (23.5°).

One 'year' on Saturn lasts about 29 Earth years. Like Jupiter, one 'day' on Saturn is much shorter than an Earth day – it is just 10.6 hours.

Saturn has (at least) 60 moons. Saturn's largest moon is Titan. Titan is the second largest moon in the solar system; Jupiter's largest moon, Ganymede, is the largest moon. Titan is the only moon in our solar system that has a thick atmosphere – which, like Earth's atmosphere, is predominantly composed of nitrogen. Saturn's atmosphere also includes two gases – methane and

methane – which have warmed Titan through the greenhouse effect. Although Titan may be too cold for life, its atmosphere does contain numerous organic molecules – which all living creatures on Earth contain. Titan's surface also has several smooth areas which may be liquid methane "lakes."

Saturn has the most prominent rings, which are actually composed of numerous chunks of ice and rocks. There are also gap moons – moons that create gaps between the rings. The rings shown in the following photo appear in false color.

1.9 Uranus

Uranus (Your-uh-nus) is the seventh planet from the sun. Uranus has a radius that is 4 times that of Earth's, while its orbital radius is 19 times that of Earth's. Uranus has 15 times as much mass as Earth. Uranus, like all of the Jovian planets, has a density (1.3 g/cc) that is small compared to the terrestrial planets.

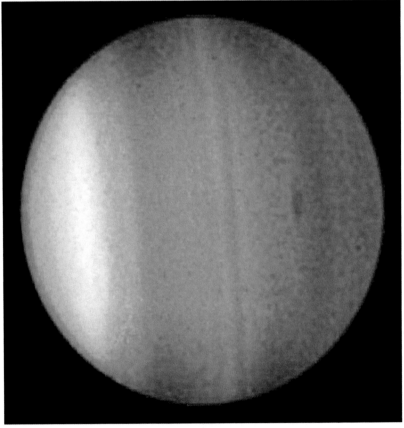

The average surface temperature of Uranus is 60 Kelvin, which is well below freezing. Like the other Jovian planets, Uranus is composed of hydrogen and helium gas.

Uranus has an extreme axis tilt of 97.9°. It rotates about an axis that is nearly perpendicular to its orbit. Its rings and the orbits of its moons are similarly tilted about 90°. This extreme tilt may have been caused by a collision during the early stages of the solar system. Uranus has the most extreme seasons of any planet in our solar system due to its unusual tilt. Because of this tilt, although Uranus has a rotational period of 17 hours (which we would normally call a 'day'), the polar regions are light and then dark for half of a Uranus 'year.' One 'year' on Uranus lasts 84 Earth years, which means it stays dark for 42 years, then light for 42 years, and so on, near the polar regions. This is what creates the extreme seasonal effects.

Uranus has (at least) 27 moons. Uranus has 5 large moons: Titania, Oberon, Umbriel, Ariel, and Miranda. Titania is the largest of Uranus's moons, and is only slightly larger than Oberon. Like Jupiter, but unlike Saturn, Uranus has a faint ring system. The following photo shows Uranus's largest moon, Titania. Miranda, the smallest of Uranus's 5 large moons, has marked features of plate tectonics (including cliffs higher than the Grand Canyon) and past geological activity.

The following photo shows Uranus with all five of its large moons. Titania (also shown in the photo above) is actually the largest of Uranus's moons, even though it looks smaller in the photo. The reason that Titania appears smaller, when in fact it is the largest moon of Uranus, in this photo is simply that it is further away from the camera than the moons Ariel and Miranda. Miranda is actually the smallest of these five moons, though it is hard to tell this by looking at the photo. Note the geologically fascinating textures of Ariel and Miranda.

1.10 Neptune

Neptune, the eighth planet from the sun, is the most distant planet now that Pluto has been classified as a dwarf planet. Neptune has a radius that is 3.9 times that of Earth's, while its orbital radius is 30 times that of Earth's. Neptune has 17 times as much mass as Earth. Neptune has the highest density (1.6 g/cc) of any Jovian planet (but is still much less dense than any terrestrial planet).

Neptune and Uranus are similar in many ways. For one, both appear blue. They also have nearly the same size and mass. Neptune is slightly smaller than Uranus, yet it is slightly more massive because it has a greater average density (recall that density equals mass divided by volume). The gap between the orbits of Uranus and Neptune is greater than the gap between the orbits of any other two consecutive planets.

The average surface temperature of Neptune is also about the same as Uranus – 60 Kelvin. Like the other Jovian planets, Neptune is composed of hydrogen and helium gas. One way that Neptune is much different from Uranus is that it has an axis tilt of 26.7° (compared to the 97.9° tilt of Uranus), which is a little more than Earth's (23.5°).

An Introduction to Basic Astronomy Concepts

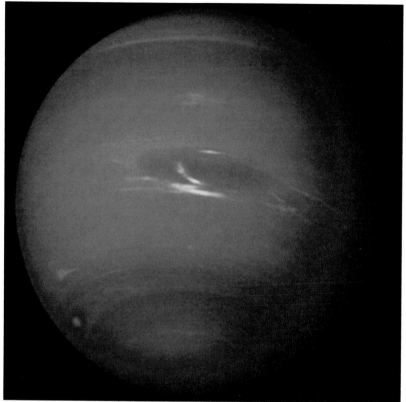

One 'year' on Neptune lasts more than one century – 165 Earth years. Like the other gas giants, one 'day' on Neptune is much shorter than an Earth day – it is just 16 hours.

Neptune has (at least) 13 moons. Neptune's largest moon is Triton. Triton is the only moon in our solar system to orbit its planet backwards compared to all of the others. Because of this strange orbit, Triton was probably captured by Neptune in its history. The following photo shows Triton. Like Jupiter and Uranus, but unlike Saturn, Neptune has a faint ring system.

49

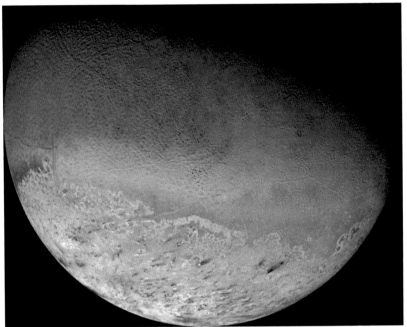

The following photo shows Neptune along with four of Neptune's other moons: Proteus, Larissa, Galatea (Gal-uh-tee-uh), and Despina. Proteus is the largest of these four moons, yet much smaller than Triton.

Since three of the largest moons in the solar system are named Titan, Titania, and Triton, it can be a challenge to keep these names straight. There is a simple way to remember this correctly. First, write the names of the largest moon of the four gas giants in alphabetical order: Gaynmede, Titan, Titania, and Triton. These moons correspond to the order of the gas giants in orbital distance from the sun: Jupiter, Saturn, Uranus, and Neptune. (Note that the names of these four planets, however, are

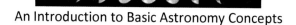
not in alphabetical order.) That is, Ganymede is Jupiter's largest moon, Titan is Saturn's largest moon, Titania is Uranus's largest moon, and Triton is Neptune's largest moon.

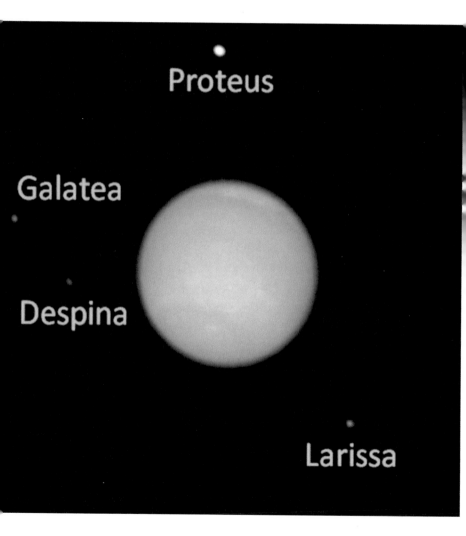

1.11 Dwarf Planets

Numerous icy bodies reside in a large belt beyond Neptune called the Kuiper (Kie-per) Belt. Those icy bodies that have an elliptical enough orbit to venture into the inner solar system and develop tails we call comets, while the other icy bodies are also comet-like in that they have similar composition. There are tens of thousands of fairly large (about 100 kilometers or more) icy bodies in the Kuiper Belt. Only the very largest of these are called dwarf planets.

The two largest known Kuiper Belt objects are the dwarf planets Eris (Ee-ris) and Pluto. Eris is actually the largest dwarf planet – Pluto is smaller. Eris and Pluto are shown in the following picture, along with Earth and its moon for size comparison. Dwarf planets are too small to see without the aid of a telescope. The images of Eris and Pluto are actually an artist's concept.

Why isn't Pluto considered to be a planet anymore? There are several reasons that astronomers recently reclassified Pluto as a dwarf planet. For one, astronomers discovered Eris, which is larger than Pluto – so they would either have to add Eris as a tenth planet, or reclassify Pluto as

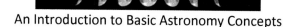

something other than a planet. Secondly, astronomers learned that Pluto and Eris are part of a belt that consists of countless icy bodies – whereas what we consider to be planets have enough mass that they clear out most of the other matter in their orbital neighborhood. Another consideration is that Pluto and Eris have icy compositions quite unlike the other outer planets, which are all gas giants – instead, they are very similar to comets and other icy objects in the Kuiper Belt. Finally, Pluto and Eris have highly elliptical orbits, and their orbits are significantly tilted relative to the plane of the solar system. Thus, it became clear to astronomers that Pluto and Eris are comet-like Kuiper Belt objects, distinctly different from planets.

Eris, the largest dwarf planet, is slightly larger than Pluto. Eris and Pluto are each about two-thirds the size of Earth's moon. Eris and Pluto each have a tiny rocky core surrounded by a thick layer of ice. Eris has (at least) one moon called Dysnomia (Dis-know-me-uh).

Pluto, the second largest known dwarf planet, has four moons. Its largest moon is named Charon (Ker-uhn). Note that Charon is not pronounced Sharon, nor is it pronounced Karen, but is correctly pronounced Ker-uhn. Pluto's orbit is so elliptical that for a decade or so out of every 248 years Pluto is actually closer to the sun than Neptune.

Five dwarf planets include Eris, Pluto, Haumea (How-may-uh), Makemake (Mah-kay-mah-kay), and Ceres (Sir-ease). Note that Makemake is not pronounced with two "keys," but with two "kays." The last, Ceres, is actually a large asteroid located in the Asteroid Belt between Mars and Jupiter, whereas the other dwarf planets are members of the Kuiper Belt beyond Neptune. We will discuss Ceres in the following section.

Beyond the Kuiper Belt lies the Oort (Ort) Cloud. The Oort Cloud is a sparse region of space that extends beyond the Kuiper Belt to the edge of our solar system. The Oort Cloud is about a 1000 times wider than the Kuiper Belt. The Oort Cloud is about 50,000 times wider than Earth's orbital radius (that's almost a light-year!), while the Kuiper Belt is about 50 times wider than Earth's orbital radius. For comparison, Neptune's orbital radius is about 30 times greater than Earth's orbital radius, while the distance to the nearest star is about 4 light-years (the distance that light travels in one year). The Oort Cloud is spherical in shape, whereas the Kuiper Belt is much more planar (like the inner solar system). The comets and other icy bodies in the Oort Cloud have more random and eccentric orbits, on average, than those in the Kuiper Belt. There are about a trillion (1,000,000,000,000) comets in the Oort Cloud.

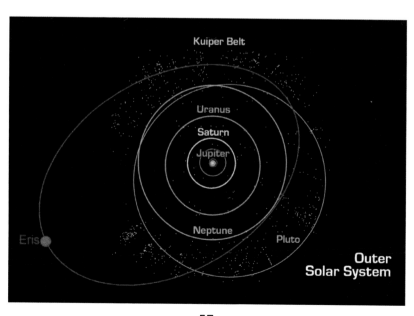

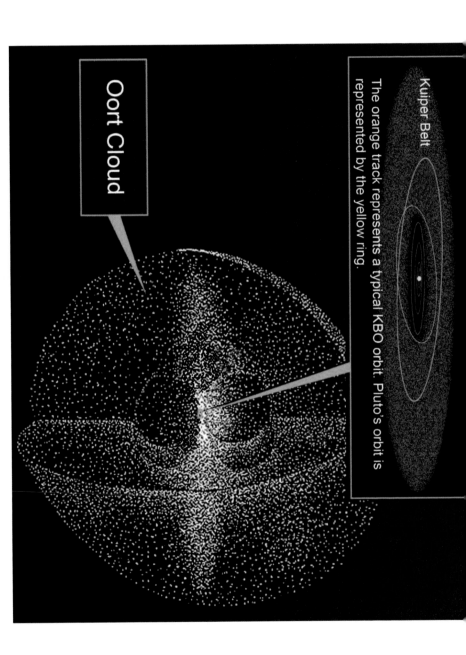

Oort Cloud

Kuiper Belt

The orange track represents a typical KBO orbit. Pluto's orbit is represented by the yellow ring.

1.12 Asteroids, Meteors, and Comets

There are countless chunks of rock that are leftovers from the formation of the planets in our solar system, which are called asteroids. Most of the asteroids orbit the sun in the Asteroid Belt between Mars and Jupiter. There are at least a hundred thousand asteroids that measure over a kilometer across, yet the combined mass of all of the asteroids in the solar system wouldn't be enough to make one small planet. The two largest known asteroids are Ceres (Sir-ease) and Vesta. The largest asteroid in the solar system, Ceres, is about 1000 kilometers in diameter (less than half the size of Pluto). Asteroids are too small to see without the aid of a telescope.

Because asteroids have much less mass than planets and large moons, they don't have enough of a gravitational pull to be round or to have a thick atmosphere. Therefore, asteroids tend to be irregularly shaped and also tend to have many craters. A few asteroids actually have their own small satellites. The following photo shows asteroid Ida and its satellite Dactyl.

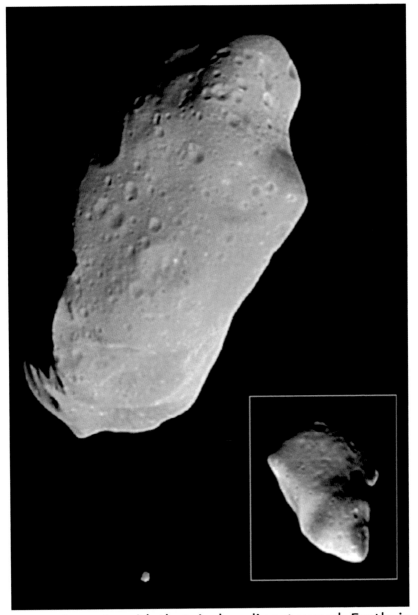

An asteroid that is heading toward Earth is called a meteoroid while it is still in space. When it enters Earth's atmosphere, it is called a meteor. We

see the meteor because it emits a flash of light as it passes through the atmosphere. If part of the meteor reaches the ground (instead of completely burning up), we call the piece that remains a meteorite. Since a meteorite was originally an asteroid, it is a small rock. The meteorite shown in the following figure was found in Antarctica in 1981 and matches rocks gathered from the moon during an Apollo mission, showing that meteoroids do not necessarily originate from the Asteroid Belt. The two faces of the block showing have the letters E and T – the abbreviation (ET) for extraterrestrial.

If the meteoroid has enough mass to leave a substantial meteorite when the meteor reaches the ground, it can result in a significant impact. The Earth has suffered a few major impacts in its history.

California

Mathilde

Lutetia

Vesta

Ceres

Pluto

Earth's moon

The following photo shows Meteor Crater in Arizona, which is over a kilometer wide and is 200 meters deep.

A comet is an icy body that travels in an elongated elliptical orbit and develops a head and tail as it passes near the sun. Numerous comets reside in the Kuiper Belt and Oort Cloud. According to Kepler's second law (that orbiting bodies sweep out equal areas in equal times), comets travel faster when they are near the sun. (Kepler's second law is described in Section 7.2.) Therefore, we only see a comet for a brief time, as it spends most of its time farther from the sun where it travels more slowly. Whereas we see light from a meteor burning up in the atmosphere, we see a comet out in space when it develops its coma and tail as it passes near the sun. Some comets return fairly frequently – such as Halley's Comet which returns every 76 years – while many comets make one pass near the sun just once in a thousand or so years. A comet is often

described as a "dirty snowball." A comet's tail always points directly away from the sun (so it often points sideways, and not backwards, compared to the direction of the comet's velocity). As the nucleus of the comet is heated as it passes near the sun, it creates a visible coma (head) – an atmosphere of gas that comes from heating the ice – and develops a tail (partly formed by gas that escapes from the coma when it is pushed by the solar wind).

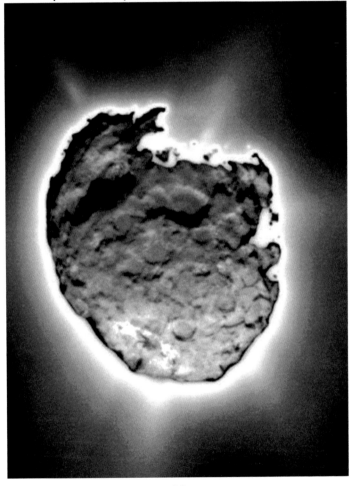

2 Understanding the Lunar Phases

2.1 Lunar Motion

A s the Earth revolves around the sun, the moon revolves around the Earth. It takes 29 and ½ days for the moon to complete its orbit around the Earth. That's about one month ("moon-th").

The Earth's orbit is only slightly elliptical; it is approximately circular. The moon's orbit is significantly more elliptical. The plane of the moon's orbit around the Earth is also tilted about 5° compared to the plane of the Earth's orbit around the sun.

The same side of the moon always faces the Earth. The side of the moon that we see from Earth is called the near side of the moon, and the other side is called the far side of the moon. If you lived on the near side of the moon, you would always see the Earth in the sky – it would never rise or set (whereas

we do see the moon rise and set in the sky from Earth). On the other hand, if you lived on the far side of the moon, you would never see the Earth.

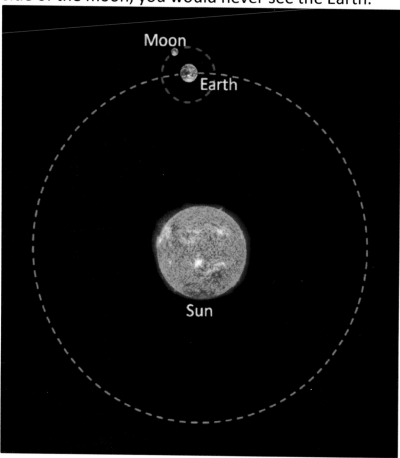

Don't confuse what we refer to as the near and far sides of the moon with the light and dark sides of the moon. If you lived on the moon, it would be light for two weeks, then dark for two weeks, then light for two weeks, and so on – regardless of whether you lived on the near or dark side of the moon.

Do you remember seeing the moon during the daytime? Do you remember not seeing the moon at night? Obviously, the moon sometimes appears during the day and sometimes appears during the night. Look for the moon every day for a few weeks, and you can easily convince yourself that this is true. The sun and the moon are the only two objects in the solar system that are visible during the daytime.

It's a common misconception, though, to associate the sun with day and the moon with night. While it is true that the sun can only be seen during the day, the moon appears sometimes during the day and sometimes does not appear at night.

2.2 The Lunar Phases

One-half of the moon is always light, and one-half of the moon is always dark. The side facing the sun is light, while the side facing away from the sun is dark. However, as the moon orbits the Earth, when we view the moon from Earth it appears to pass through different phases.

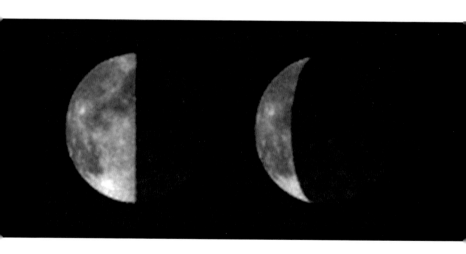

The phases of the moon are caused by the ever-changing positions of the moon and sun relative to the Earth. They are <u>not</u> caused by shadows of the Earth cast on the moon. We will demonstrate this through a variety of diagrams in later sections.

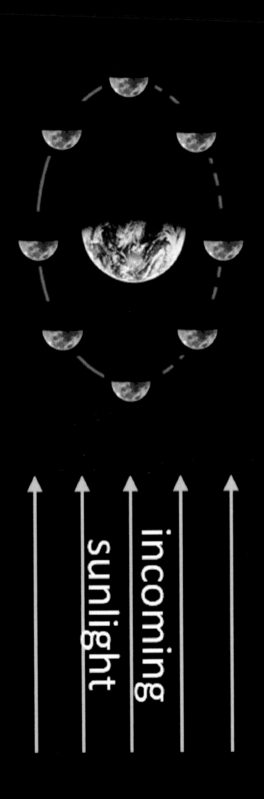

During the new moon phase, no reflected sunlight is visible; the moon looks completely dark. Look closely at the next photo to find the moon.

During a crescent phase, reflected sunlight appears in less than one-half of the moon's visible surface.

During a quarter moon, one-half of the moon's visible surface appears to be lit by sunlight. It's important to note that we call the following photo

(<u>not</u> the previous photo) a quarter, even though we see half of the moon. The reason that we call it a quarter is that the moon grows from a new moon into a quarter in one-fourth of the lunar cycle. In the next section, we will see that there are actually two different quarters – a first and third quarter. (Second quarter is full, fourth quarter is new.)

During a gibbous moon, reflected sunlight appears in more than one-half of the moon's visible surface.

During a full moon, reflected sunlight appears in the shape of a full circle.

For one-half of a lunar cycle, the moon waxes – i.e. it grows from a new moon to a full moon. For the other half of a lunar cycle, the moon wanes – i.e. it transforms from a full moon to a new moon.

The moon is waxing when the amount of reflected sunlight seen is increasing. The moon is a waxing crescent between the new moon and first quarter phases and a waxing gibbous between the first quarter and full moon phases. The moon is waning when the amount of reflected sunlight seen is decreasing. The moon is a waning gibbous between the full moon and third quarter phases and a waning crescent between the third quarter and new moon phases.

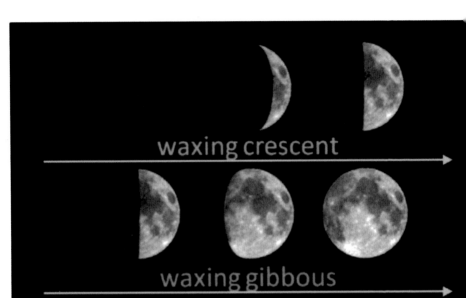

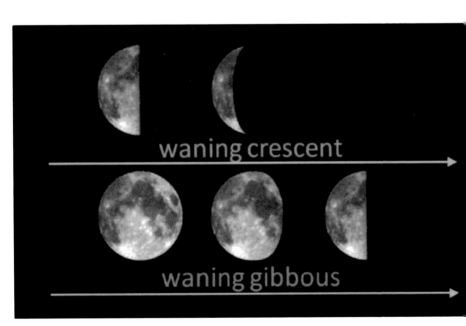

2.3 The Lunar Cycle

Let's look at one lunar cycle in order from a new moon to a full moon and back to a new moon. We begin with the new moon.

The moon then grows, so next is a waxing crescent.

The waxing crescent turns into the first quarter. It looks like half a moon, but we call it a quarter because it takes one-fourth of a lunar cycle for the moon to wax into the first quarter starting with a new moon.

The moon continues to grow, so next is a waxing gibbous.

We reach a full moon when it is finished waxing. Watch out for those werewolves!

Now the moon wanes, so next is a waning gibbous.

The waning gibbous turns into the third quarter. We call it the third quarter because it takes three-fourths of the lunar cycle to reach the third quarter starting from a new moon.

The moon continues to wane, so next is a waning crescent.

The lunar cycle is complete when it returns to a new moon.

All of the lunar phases in the lunar cycle are shown in the following photo. Read the photo from left to right, top to bottom (just as you would normally read a book).

Lunar Phases

new moon

waxing crescent

first quarter

waxing gibbous

full moon

waning gibbous

third quarter

waning crescent

new moon

2.4 How the Quarter Moon Forms

In the diagram on the following page, a quarter moon is seen from Earth. Sunlight illuminates the moon, and reflected light from the moon is viewed from Earth.

In the diagram on the following page, from Earth the right half of the moon appears lit, while the left half of the moon appears dark. To see this, imagine standing at point A and looking toward the moon. The moon would look like this:

2.5 How the Full Moon Forms

In the following diagram, a full moon is seen from Earth. Follow the sunlight as it illuminates the moon and reflects toward the Earth.

In the diagram on the following page, from Earth the entire moon appears lit. To see this, imagine standing at point A and looking toward the moon. It looks like it should be a lunar eclipse, but is not because the moon, Earth, and sun do not lie in the same plane when a full moon is formed (to be explained when we discuss how eclipses are formed in Chapter 3). The moon would look like this:

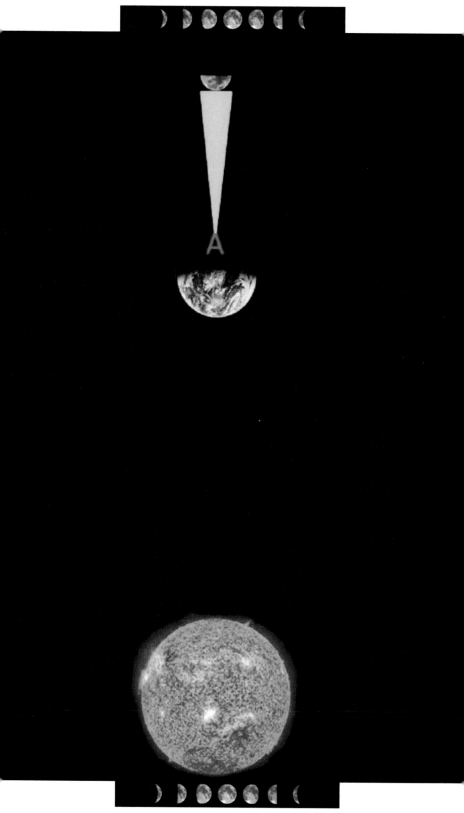

2.6 How the New Moon Forms

In the following diagram, a new moon is seen from Earth. Follow the sunlight as it illuminates the moon and reflects and you will see, that in this position, no sunlight reflects from the moon toward the Earth.

In the diagram on the following page, from the Earth none of the moon appears lit. To see this, imagine standing at point A and looking toward the moon. It looks like it should be a solar eclipse, but is not because the moon, Earth, and sun do not lie in the same plane when a new moon is formed (to be explained when we discuss how eclipses are formed in Chapter 3). The moon would look like this:

2.7 How the Crescent Moon Forms

In the following diagram, a crescent moon is seen from Earth. Follow the sunlight as it illuminates the moon and reflects toward the Earth. In this position, less than half of the moon appears to be lit from the vantage point of Earth.

To see this, imagine standing at point A and looking toward the moon. Less than half of the moon would be lit; you would just see a sliver on one side. The moon would look like this:

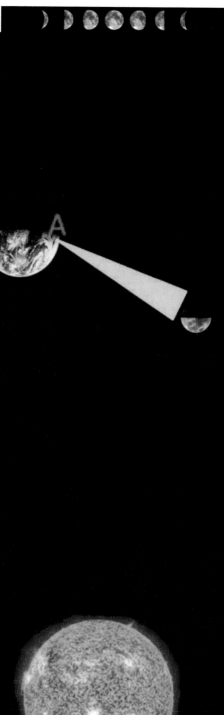

2.8 How the Gibbous Moon Forms

In the following diagram, a gibbous moon is seen from Earth. Follow the sunlight as it illuminates the moon and reflects toward the Earth. In this position, more than half of the moon appears to be lit from the vantage point of Earth.

To see this, imagine standing at point A and looking toward the moon. All of the moon appears lit except for a sliver on one side. The gibbous phase is an inversion of the crescent phase: The dark part of a gibbous looks like a crescent. The moon would look like this:

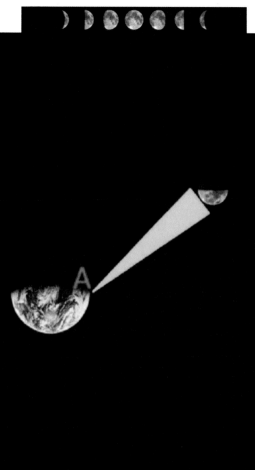

2.9 The Lunar Phases in the Moon's Orbit

The figure on the following page illustrates how the different phases of the moon – as viewed from Earth – are formed. It may help to really study and think about this diagram, and to compare it with the diagrams from the previous sections.

The large moons show the moon's orbit around the Earth. The small moons show the phase of the moon as viewed from Earth when the moon is in that position.

It has to do with the relative positions of the Earth, moon, and sun. (It is incorrect to think that the phases of the moon are shadows formed by sunlight that is blocked by the Earth.)

One-half of the moon is always illuminated by sunlight, but from Earth we see every phase from a new moon to a full moon depending on where the moon is in its orbit around the Earth.

Below, all of the phases of the lunar cycle appear in order (from left to right) every 29.5 days.

It takes 29.5 days for the moon to complete its orbit around the Earth.

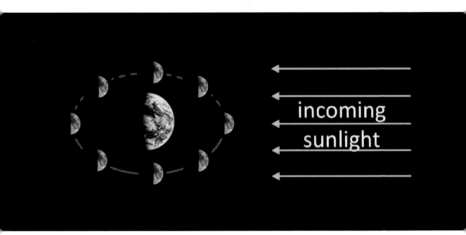

After one-fourth and three-fourths of the lunar cycle, the moon appears as a half-moon. Hence the names first and third quarter. (The moon is full at second quarter and new at fourth quarter.)

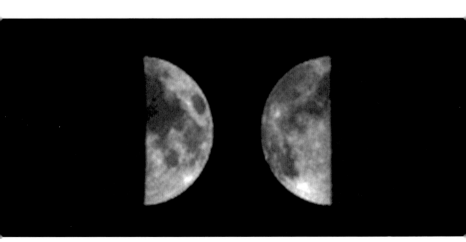

3 Understanding Solar and Lunar Eclipses

3.1 How a Lunar Eclipse Forms

A lunar eclipse occurs when the Earth blocks sunlight from reaching the moon. When this happens, the Earth casts a shadow on the moon's surface. A lunar eclipse is illustrated on the following page.

A lunar eclipse can only occur during a full moon. However, during most full moons, there is no lunar eclipse (as we will explore in Section 3.3) – or there is only a partial lunar eclipse.

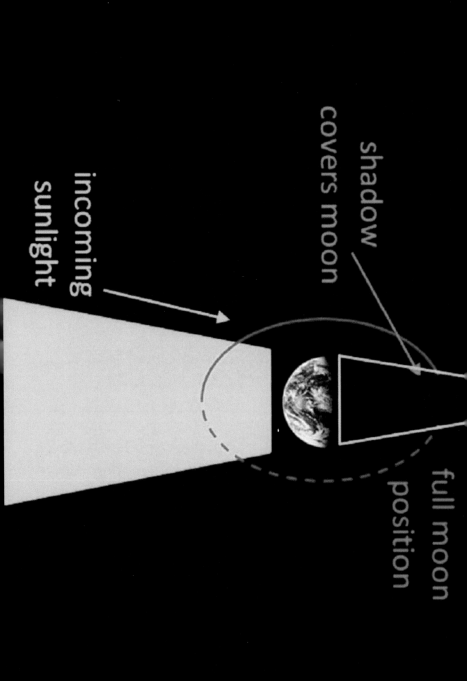

incoming
sunlight

shadow
covers moon

full moon
position

3.2 How a Solar Eclipse Forms

A solar eclipse occurs when the moon blocks sunlight from reaching Earth. When this happens, the moon casts a shadow on Earth's surface. A solar eclipse is illustrated on the following page.

A solar eclipse can only occur during a new moon. However, during most new moons, there is no solar eclipse (as we will explore in Section 3.3).

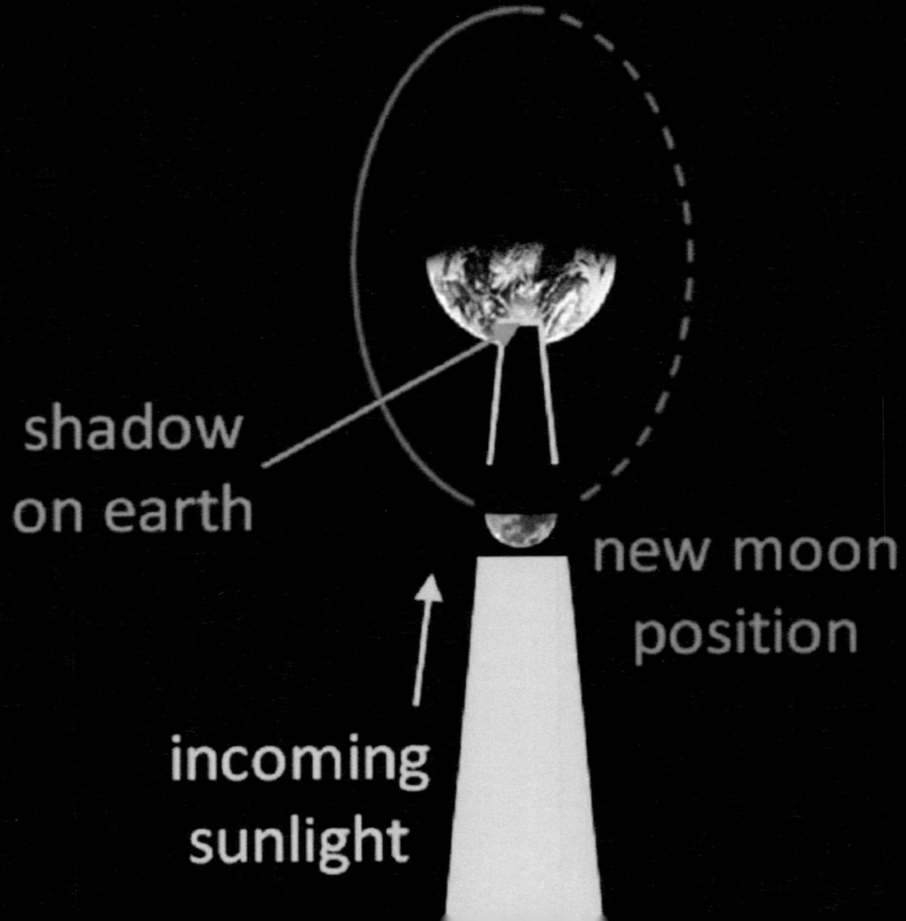

shadow on earth

incoming sunlight

new moon position

3.3 Why Eclipses Are Rare

We see a full moon and new moon every month, but solar and lunar eclipses are rare.

The reason that eclipses are rare is that the plane of the moon's orbit around the Earth is tilted about 5° compared to the plane of the Earth's orbit around the sun. An eclipse can only occur when the moon happens to lie near the plane of the Earth's orbit around the sun.

The following diagram illustrates the Earth in four different positions in its orbit around the sun. The moon is also shown in both the new and full moon positions for each position of the Earth. (However, the moon is not shown as a full or new moon because we are not viewing it from Earth's perspective in the diagram. Go by the wording – new or full – and not by how the moon looks in the diagram from the picture's view.)

In two of Earth's positions, the Earth, moon, and sun lie in the same plane. These are the two positions in the Earth's orbit where a full eclipse can occur.

In the other two positions of the Earth shown in the diagram, the moon lies well above or below

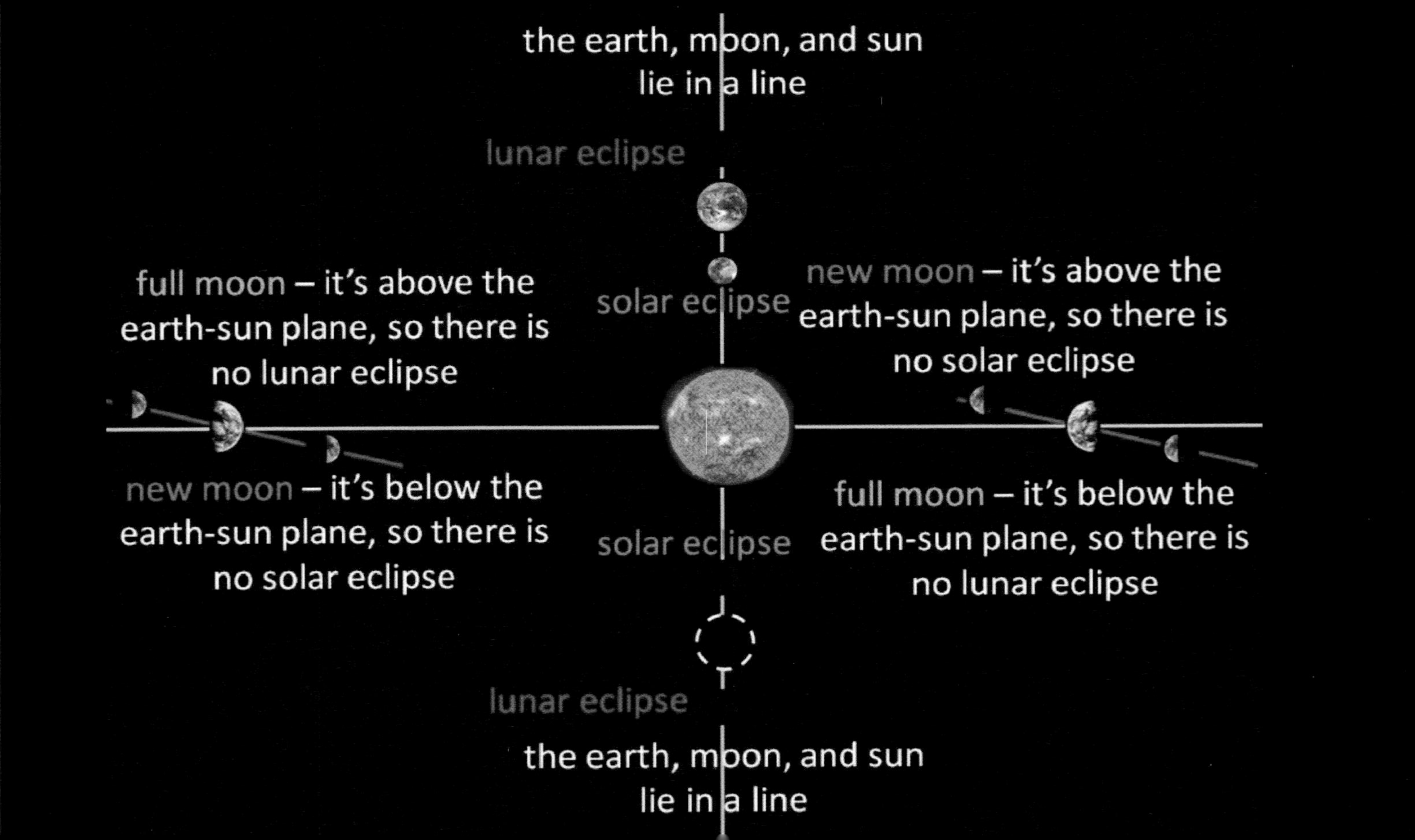

the earth, moon, and sun
lie in a line

full moon — it's above the
earth-sun plane, so there is
no lunar eclipse

new moon — it's below the
earth-sun plane, so there is
no solar eclipse

lunar eclipse

solar eclipse

new moon — it's above the
earth-sun plane, so there is
no solar eclipse

solar eclipse

full moon — it's below the
earth-sun plane, so there is
no lunar eclipse

lunar eclipse

the earth, moon, and sun
lie in a line

the Earth-sun plane (because its orbit is tilted 5° relative to Earth's orbit). No eclipse can occur in these positions.

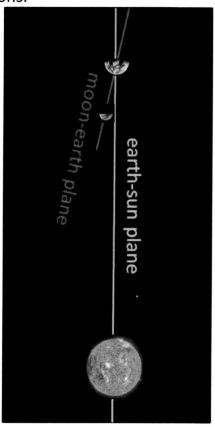

Eclipses are rare because the Earth has to be in one of two positions in its orbit (most of the year it is not near those two positions) and there must be a new or full moon when the Earth is in these positions. When a solar eclipse does occur, the moon only blocks sunlight over a small area on Earth's surface, which makes it even rarer to observe a solar eclipse.

4 Understanding the Seasons

4.1 The Intensity of Sunlight

The intensity of sunlight equals power per unit area. Power is the rate at which the sun's energy is radiated outward. The sun's power is constant – it radiates energy at a constant rate. However, the intensity of sunlight varies with distance from the sun, and also depends on the angle of the sun's rays.

Let us first consider how the intensity of sunlight depends on distance from the sun. Sunlight radiates outward from the sun. If you imagine a large sphere centered about the sun, all of the sunlight reaching the surface of the imaginary sphere was emitted by the sun at the same instant. As the sunlight propagates outward, this imaginary sphere grows larger and larger.

The power of the sunlight is the same regardless of the size of the sphere, but the intensity of the sunlight depends on the size of the sphere according to the formula that follows this paragraph. The area needed here is the surface area of the sphere, which is $A = 4\pi R^2$, where R is the radius of the sphere. In this case, the intensity of sunlight equals:

$$I = \frac{P}{4\pi R^2}$$

where I is the symbol for intensity and P is the symbol for power.

According to this formula, the intensity of sunlight gets smaller as the sunlight gets farther from the sun. This should make sense, conceptually. The formula explains why sunlight is much more intense on Mercury, which is the closest planet to the sun, and much weaker on Neptune, which is the furthest planet from the sun.

This concept is illustrated visually in the following example. The diagram shows several rays propagating outward from the sun. The same number of rays reach the orbits of Mars and Earth in this picture because the sun's power is constant, but the intensity of sunlight is greater on Earth than Mars because the sun's rays are more spread out at Mars's orbit than at Earth's orbit. The rays are distributed over a greater area at Mars's orbit compared to Earth's orbit (remember, it's three-dimensional, so what look like circles in the diagram are really spheres), which diminishes the intensity of sunlight at Mars compared to Earth. (Of course, there are more rays of sunlight than shown here, but this illustrates the basic idea.)

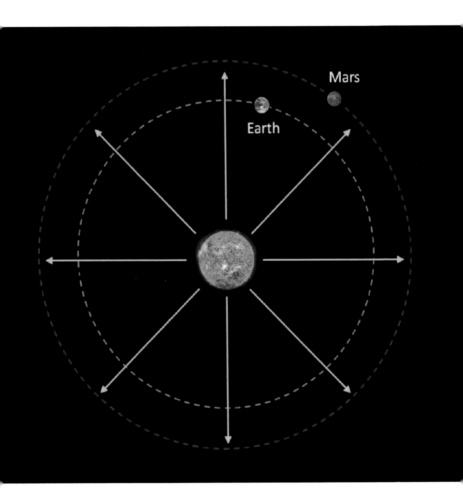

4.2 The Direction of the Sun's Rays

The previous example, with the growing spheres depicting sunlight traveling radially outward from the sun, explains how intensity depends on distance from the sun. Let us now consider how the intensity of sunlight depends on the angle of the sun's rays. The angle of the sun's rays will be important for understanding why there are seasons on Earth.

When sunlight is perpendicular to a surface, the rays are more concentrated. When sunlight strikes a surface at an angle, the rays are more spread out. Sunlight is more intense when the rays are more concentrated, and less intense when the rays are more spread out. (That's why you can use a lens to set something on fire – the lens focuses the rays, making the sunlight much more intense near the focal point of the lens.)

The two images on the right side of the following diagram show that the rays of sunlight are more spread out when they strike a surface at an angle. The two images on the left side show that the sunlight is more direct during the summer and strikes the surface at an angle during the winter.

More rays of sunshine reach a given area of the northern hemisphere in the summer; fewer rays reach a given area in winter. This is illustrated in the following diagram. It may be useful to compare the following diagram with the diagram of Section 4.3.

December

incoming sunlight

fewer rays reach a given area

2 points further apart (more area between 2 rays)

June

incoming sunlight

2 points closer together (less area between 2 rays)

more rays reach a given area

4.3 The Cause of Earth's Seasons

Two things affect the intensity of the sun's rays. In Section 4.1 we learned that the intensity of sunlight depends on distance from the sun, and in Section 4.2 we learned that the intensity of sunlight also depends on the direction of the sun's rays. However, only one of these two effects causes Earth's seasons.

The distance from the sun and Earth does <u>not</u> have a significant effect on the seasons. Some students who learn that Earth orbits the sun in an ellipse intuitively expect the sun to be closer to the Earth in the summer and further from the Earth in the winter. However, this is <u>not</u> true. Although the Earth does orbit the sun in an ellipse, the Earth's orbit is only slightly elliptical — it is very nearly circular. Therefore, the Earth is not significantly closer or further from the sun over the course of its orbit for this to have a significant effect on the intensity of sunlight received by the Earth.

There is a simple argument that can convince you that variations in the Earth-sun distance are not important for seasonal effects. You may know that it is wintertime in the southern hemisphere when it is

summertime in the northern hemisphere, and vice-versa. For example, when it's summer in the United States, it's winter in Australia. So when we experience summer, one-half of the Earth is simultaneously experiencing winter. It's both winter and summer on the Earth twice every year. Obviously, then, it's not distance that affects the seasons. Whenever it's summer somewhere on the Earth, it's also winter somewhere on the Earth. Changes in the Earth-sun distance don't explain this.

What is different between the northern and southern hemispheres? It's the angle of the sun's rays due to the tilt of Earth's axis of rotation.

The tilt of Earth's axis (<u>not</u> the Earth-sun distance) determines the seasons.

Sunlight shines more directly on the northern hemisphere in the summer, so more light reaches a given area of the northern hemisphere's surface; in winter, sunlight enters at an angle such that less light reaches a given area.

You can see this in the following diagram. When it is June, the northern hemisphere is tilted toward the sun. The sun's rays strike the northern hemisphere more directly, providing a greater intensity of sunlight, causing summer. In June, the southern hemisphere is tilted away from the sun. The sun's rays strike the southern hemisphere less directly, providing less intensity of sunlight, causing winter. It is just the opposite in December.

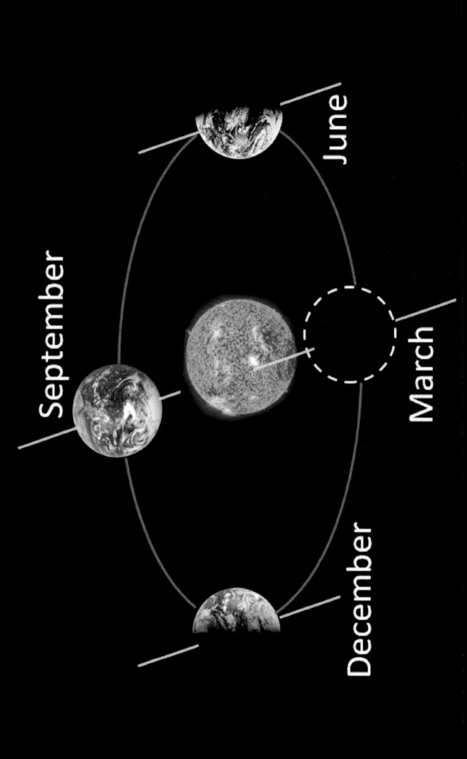

5 Evidence that the Earth is Round

5.1 Evidence from Sailing

For several centuries, sailors have observed that the Earth is round. Watch how the boat disappears differently in the following two diagrams. We will call these Figure 1 and Figure 2.

Figures 1 and 2 are illustrated in the diagrams on the following pages. In Figure 1, Observer A, on the cliff, can see the top part of the ship. However, Observer B, on the beach, can't see the ship at all.

In Figure 2, both Observers A and B see all of the ship. The ship appears smaller as it leaves, but the same proportion of the ship is always visible.

We can use these figures to determine if the Earth is round or flat. We use two observers, just like the figures. One observer stands on top of a cliff near the ocean, while the other observer stands on the beach below. The results of this experiment show, conclusively, that the Earth is round.

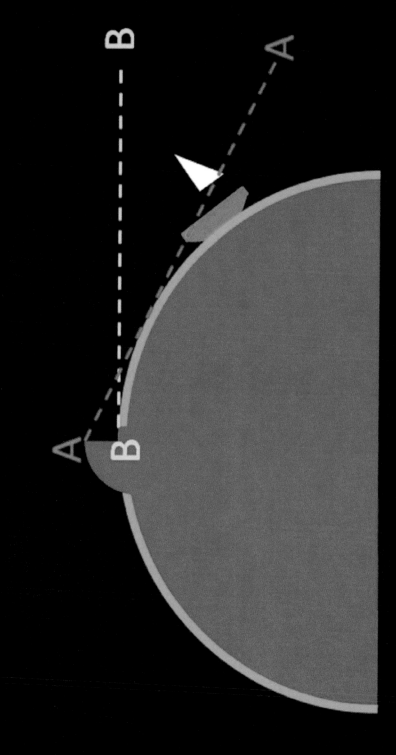

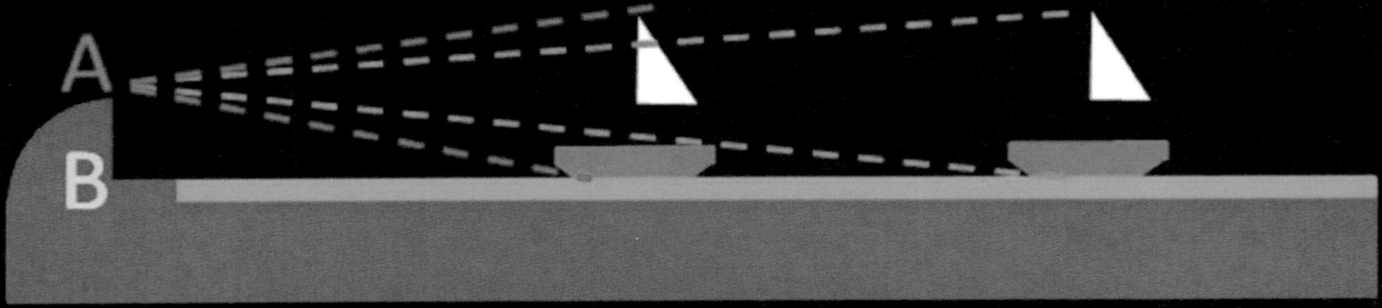
A
B

When the observer on the beach (Observer B) announces that he can no longer see the ship, the observer on the cliff (Observer A) will still be able to see the ship for some time. Also, the observers will both notice that when the boat disappears, the bottom of it disappears first. If the Earth were flat, both observers should instead see the boat disappear at the same time, and should always be able to see the whole boat, instead of seeing the bottom disappear first. This is one way that we know that the Earth is round.

5.2 Evidence from Sunlight

Eratosthenes (Air-uh-toss-thuh-knees) was a Greek mathematician who measured Earth's circumference circa 235 B.C. while serving as the chief librarian in Alexandria, Egypt.

During the summer solstice on June 22, at high noon, a vertical pillar cast a shadow in Alexandria, yet at the same time, a vertical pillar cast no shadow in Syene (Sigh-ee-knee). Eratosthenes understood that this discrepancy was caused by the shape of the Earth being round instead of flat. He made some measurements and used them to calculate the size of the Earth, knowing that the Earth was spherical in shape.

The following diagram illustrates two towers – one in Syene and another in Alexandria. Both towers are vertical, and in both cases the ground below is level. The yellow tower in Syene casts no shadow, while the pink tower in Alexandria casts a shadow indicated by the dashed blue line. The solid blue curve represents the shape of the Earth. In this diagram, we can see how the shape of the Earth explains why one vertical tower in Syene could have no shadow while another vertical tower in Alexandria has a shadow. This would not happen if the Earth were flat.

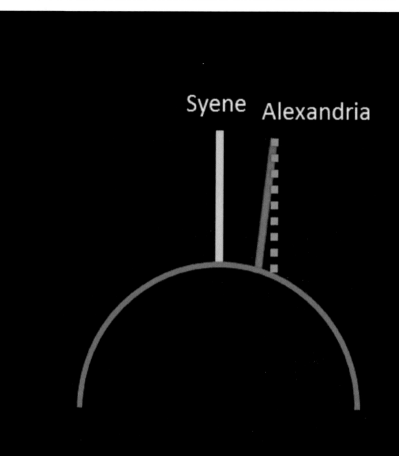

Knowing the distance between Alexandria and Syene, Eratosthenes used geometry to calculate Earth's circumference. He was within 5% of the modern value of 40,000 km (or 40,000,000 m). Driving 60 mph (that's about 100 km/hr), it would take 400 hours (nearly 17 days nonstop) to drive around Earth's circumference (going 'straight'!) – but first you better build the highway over the oceans.

⚙ Here is the calculation of earth's circumference, C:

$$\frac{\theta}{2\pi} = \frac{d}{C}$$

distance between cities, d

earth's radius

Syene   Alexandria

height

shadow

$$\tan\theta = \frac{shadow}{height}$$

θ

equal angles

5.3 Evidence from Spacecraft

Nowadays, we have these convincing space photos from NASA, like the one below. Since we have launched numerous satellites, rockets, and spaceships into space, several astronauts have seen Earth from space and several cameras have photographed Earth from space. This is very clear evidence that the Earth is roughly spherical in shape.

6 Models of Our Solar System

6.1 The Geocentric Model

Ptolemy (Tall-uh-me) was a Greek mathematician and librarian who developed a model of our solar system with Earth at the center circa 140 A.D. This is called the geocentric model (also known as the Ptolemaic model). (The "gee" refers to Earth, as in geology.)

The geocentric model has the Earth in the center of the solar system. The sun orbits the Earth in this model between the orbits of Venus and Mars – it separates the inner planets from the outer planets.

We call Mercury and Venus the inner planets because we know that the sun really lies at the center with the Earth orbiting between Venus and Mars, while we call Mars, Jupiter, Saturn, Uranus, and Neptune the outer planets. In the geocentric model, Mercury and Venus are inner planets in the

sense that they were inside the sun's orbit, and the others were still outer planets compared to the sun's orbit.

Mercury and Venus behave much differently in the night sky than Mars, Jupiter, Saturn, Uranus, and Neptune – so even the ancient Greeks realized that there were two distinctly different types of planets (in terms of their motion relative to the Earth).

Mercury and Venus are usually seen close to the horizon when they are visible from Earth, and are usually seen shortly after sunset or shortly before sunrise. You won't find either directly overhead during the middle of the night. However, the outer planets don't have such a restriction.

The purple pie slice in the following diagram shows the region of the geocentric model where Mercury and Venus can potentially be found. This is the geocentric model's way of accounting for why Mercury and Venus never stray far from the sun.

A strange feature in the geocentric model is that Venus is shown closer to the sun than Mercury. The ancient Greeks didn't have measurements for the distances between Mercury, Venus, and the sun, and so this original model happens to have Venus closer to the sun than Mercury. As it turns out, that's not the main problem with the model (and if that had been the only problem, it would have been very simple to correct).

The Geocentric Model

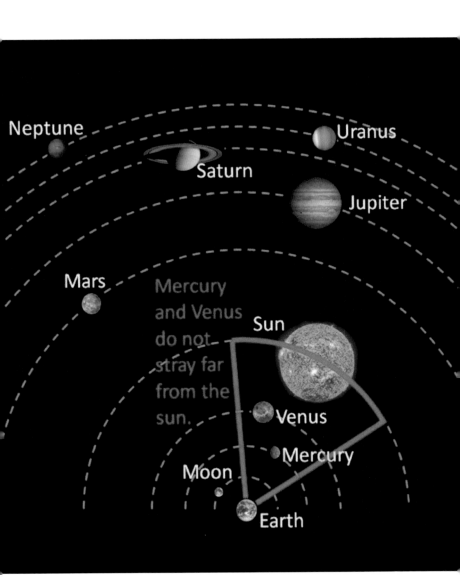

6.2 The Heliocentric Model

Aristarchus (Are-iss-tar-cuss) was a Greek mathematician who developed a model of our solar system with sun at the center circa 240 B.C. This is called the heliocentric model. (The Greek word "helios" means the sun.) Nicholas Copernicus (Coe-purr-ni-cuss), who lived from 1473-1543 A.D., revived and improved the heliocentric model.

The sun lies at the center of the heliocentric model, while the Earth orbits the sun in an orbit between Venus and Mars. The heliocentric model is simpler in that it does not require an artificial restriction of the motions of Mercury and Venus. We can naturally explain why Mercury and Venus are only visible near the sun in the heliocentric model. We will return to this point in a later section. There are also a couple of other ways in which the heliocentric model is simpler than the geocentric model, and we will also explore these ideas in a later section. Right now, the important thing is to compare the diagrams for the geocentric model (in the previous section) and the heliocentric model (in the following figure).

It's especially important to note that geocentric means Earth-centered, whereas heliocentric means sun-centered.

The Heliocentric Model

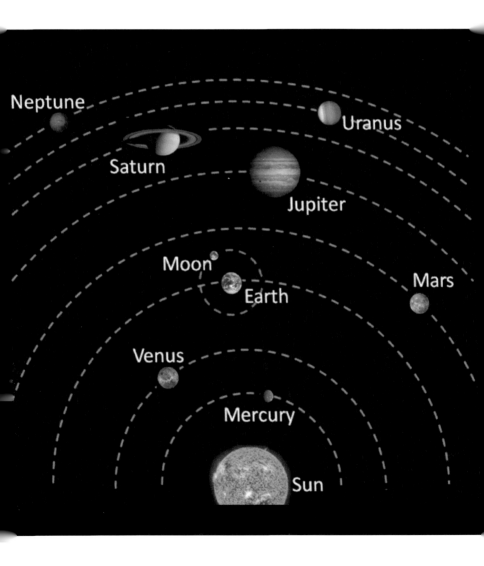

6.3 Retrograde Motion

The outer planets usually appear to move in a single direction across the night sky, as viewed from Earth. This explains why the word planet derives from the Greek word for wanderer. The planets appear to wander across the night sky.

However, when the Earth passes an outer planet, for a short period the outer planet appears to move backward. This phenomenon is called retrograde motion.

The following figure illustrates this behavior as explained in the heliocentric model. The top right figure shows the retrograde loop observed from Earth relative to the fixed stars; the main figure shows Earth viewing Mars as Earth passes it.

The geocentric explanation for retrograde motion is illustrated in the second figure that follows. The early Greeks introduced the deferent (large circle) and epicycle (small circle) in order to explain retrograde motion.

The planet rotates with the epicycle as the epicycle travels along the deferent. You might note that the heliocentric model explains retrograde motion more naturally (the heliocentric model

Retrograde motion in the heliocentric model

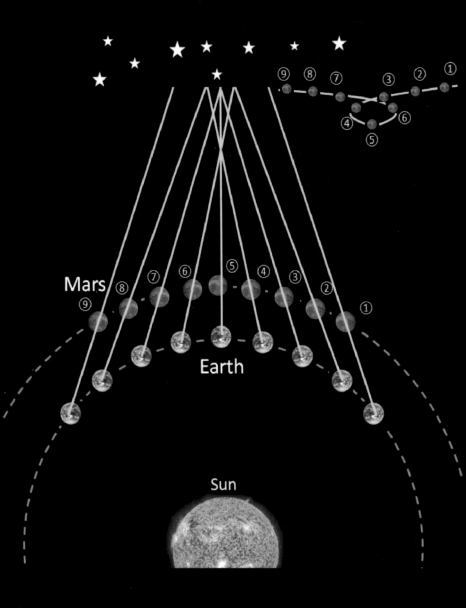

doesn't have to introduce circles within circles – just one set of circles will do). Eventually, measurements became so precise that they had to add additional epicycles in order to make the geocentric model agree with experimental observations. In this way, the geocentric model became increasingly complicated.

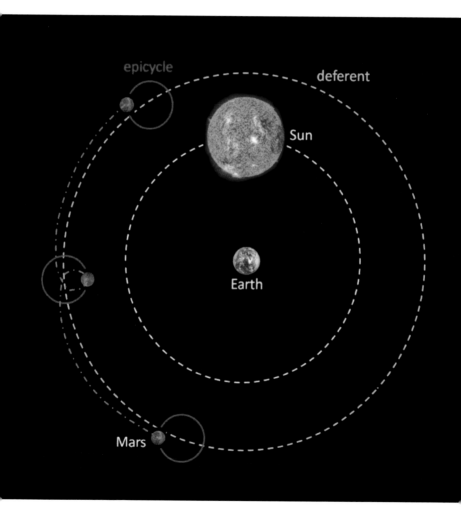

6.4 Objections to the Heliocentric Model

The Greeks raised some good objections to the heliocentric model:

(1) If the Earth were going around the sun, it would have to move incredibly fast. Why don't all loose objects, like birds and rocks, fall off of the Earth? If you run really fast, a loose cap will fall off of your head.

(2) If the Earth were orbiting the sun, it would be an enormous distance from its original position after six months. There should be a parallax shift of nearby stars compared to distant stars as a result of this change in position. Why isn't a very noticeable shift in the positions of the stars observed?

The orbital speed of the Earth relative to the sun can be found by dividing Earth's circumference $(C = 2\pi R)$ by its orbital period (one year). The answer, 30 kilometers per second, is like going from Los Angeles to New York City in 2 minutes!

Although Aristarchus argued that his heliocentric model offered a much simpler explanation for the solar system than the geocentric model, the Greeks continued to believe that the solar system was geocentric rather than heliocentric because of these objections that were raised.

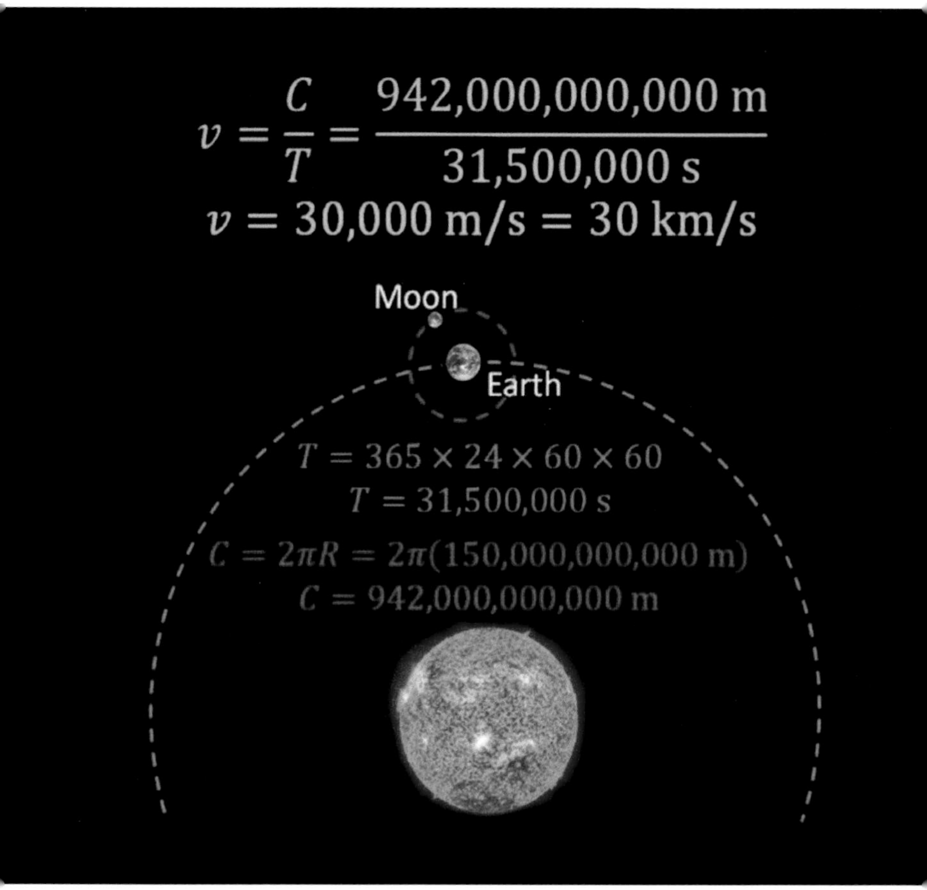

$$v = \frac{C}{T} = \frac{942,000,000,000 \text{ m}}{31,500,000 \text{ s}}$$
$$v = 30,000 \text{ m/s} = 30 \text{ km/s}$$

Moon

Earth

$$T = 365 \times 24 \times 60 \times 60$$
$$T = 31,500,000 \text{ s}$$
$$C = 2\pi R = 2\pi(150,000,000,000 \text{ m})$$
$$C = 942,000,000,000 \text{ m}$$

It wasn't until the Renaissance period that scientists were able to develop good arguments (supported by experiment) to overcome the Greek objections to the heliocentric model. We will consider how they overcame these two Greek objections to the heliocentric model in the next two sections.

6.5 The Discovery of Inertia

Imagine an airplane flying from Los Angeles to New York City with a box of bananas on its roof. Imagine that the airplane travels with a constant velocity of 30 km/s, so it gets there in about 2 minutes. Do you think that the box of bananas would fall off? Why?

The Earth is basically doing the same thing as it orbits the sun, except for traveling in an ellipse instead of a straight line. What's different about the Earth's motion?

For one, the Earth has a significant gravitational field, which attracts all massive objects. Earth is pulling rocks, birds, and all other terrestrial objects toward its center. But imagine the force it would take to prevent a box of bananas from coming off the roof of that airplane at such a high speed!

For another, there is no air resistance. The Earth travels through space. There is no air pushing against it, like there is on the airplane.

Lastly, Newton's law of inertia explains why objects don't fall off of the Earth. Isaac Newton (Newt-'n) discovered inertia in the 1600's.

All objects have inertia, which is a natural tendency to maintain constant velocity.

Velocity is a combination of speed and direction. Speed means how fast, velocity means how fast and which way.

All of the objects on the Earth have a natural tendency to keep up with the Earth. They don't have a natural tendency to fall behind. They don't fall off because they have inertia.

The box of bananas actually wants to stay on the roof of the airplane because it has inertia. However, the box of bananas would fall off of such a high-speed plane trip because of air resistance.

The Greeks incorrectly thought that objects had a natural tendency to come to rest. Isaac Newton discovered that moving objects actually have a natural tendency to keep moving. Terrestrial objects come to rest because the forces of air resistance and friction overcome their natural tendency.

Space is a near-perfect vacuum – it is completely devoid of matter. There is no air or anything else to push objects off the Earth. This is why the Earth does not slow down, but maintains its speed. Inertia and gravity explain why objects do not fall off of the fast-moving Earth in the heliocentric model.

6.6 Stellar Parallax

The following picture shows the Earth in two positions six months apart. Due to Earth's change in position, the position of a nearby star appears to shift relative to more distant stars. This phenomenon is known as stellar parallax.

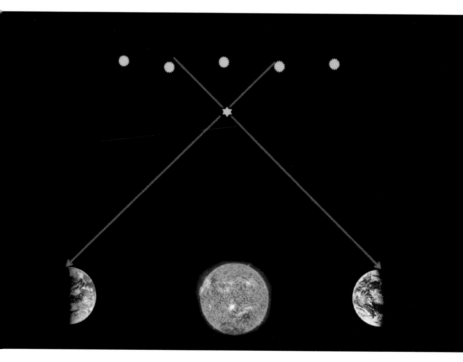

You can observe the phenomenon of parallax by placing a finger a foot before your eyes. Close one eye, then the other, and your finger will appear to shift positions.

Stellar parallax is actually a very subtle effect, unlike the previous picture, because the Earth-sun distance is very tiny compared to the distance to any stars. Stellar parallax can be observed as a very tiny shift, but it requires precise measurements with telescopes. The Greeks did not observe stellar parallax because they did not have any telescopes with which to observe this effect. Since they didn't observe the effect with their eyes, they believed that the effect didn't occur at all.

The Greeks didn't imagine that the nearest star could actually be four light-years away. Thus, when they didn't observe a noticeable stellar parallax, they incorrectly concluded that the Earth doesn't orbit the sun.

Stellar parallax actually does occur, and can be detected with the aid of a telescope, which supports the heliocentric model. The observation of stellar parallax does not fit the geocentric model.

6.7 The Modern Heliocentric Model

Following is a more modern view of the heliocentric model, which addresses the early Greek objections to it:

- The apparent motion of the sun across Earth's sky can be explained by Earth's rotation on its axis, with the sun stationary.
- The fact that Mercury and Venus can only be seen near the horizon is naturally explained in the heliocentric model: The inner planets can't stray far from the sun.
- The epicycle motion is not needed to explain retrograde motion. This was an unnecessary complication introduced to try to preserve the incorrect geocentric model.
- Isaac Newton's discoveries of inertia and gravity explain why objects do not fall off of the fast-moving Earth in its orbit around the sun.
- The stars are so far away that stellar parallax is too small to notice without a telescope.

The Heliocentric Model

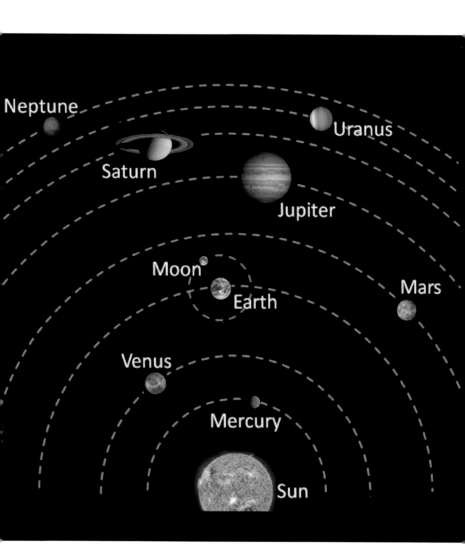

6.8 Evidence for the Heliocentric Model

Tycho Brahe (Tea-co Bra-uh), who lived from 1546 to 1601 A.D., built an observatory and made the most accurate astronomical measurements of his day. Tycho was later assisted by Johannes Kepler (Yo-hahn-us Kep-ler), who lived from 1571 to 1630 A.D.

When Kepler compared the astronomical data to Ptolemy's geocentric model and Copernicus' heliocentric model, he realized that (1) the geocentric model is incorrect and (2) the heliocentric model needed to be revised in order to explain the data.

The necessary revision was to allow the planets to travel in elliptical orbits, rather than circles, with the sun lying at one focus.

Mercury and Venus are only seen near the horizon from Earth. In order to agree with this observation, Mercury and Venus are restricted to lie near the sun in the geocentric model. Venus can only be new or a crescent in the geocentric model. Venus can't be a gibbous or full in the geocentric model.

A full Venus is not possible in the geocentric model. Only a new Venus and crescent Venus can be observed in the geocentric model, which is contrary to experimental observations. This is clear evidence that the geocentric model is incorrect.

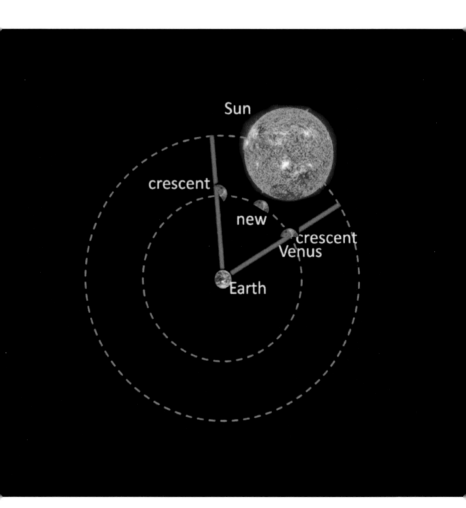

A full Venus is possible in the heliocentric model. Every phase of Venus from new to full can be observed in the heliocentric model, which is consistent with experimental observations. Observations of Venus made with telescopes support the heliocentric model.

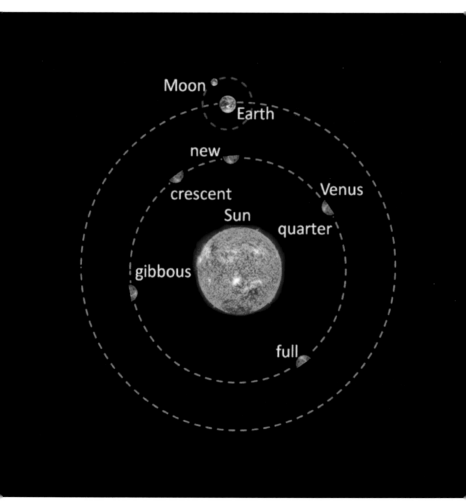

All of the phases of Venus, from new to full, are possible in the heliocentric model. Galileo (Gal-uh-lay-oh) Galilee, who lived from 1564-1642 A.D., observed a full phase of Venus through a telescope. This evidence supports the heliocentric model and rules out the geocentric model.

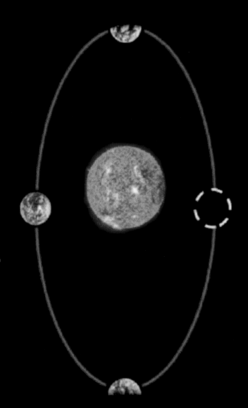

The Earth orbits the sun.
Our solar system is heliocentric.

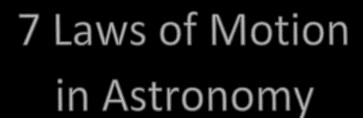

7 Laws of Motion in Astronomy

7.1 Kepler's First Law

According to Kepler's first law, the planets orbit the sun along the path of an ellipse, with the sun lying at one focus.

Kepler's first law also applies to comets and asteroids orbiting the sun, as well as moons and man-made satellites orbiting planets.

The point in a planet's orbit that is nearest to the sun is called perihelion, while the point furthest from the sun is called aphelion. (The prefix "peri" means near, as in periscope; the prefixes "ap" and "ab" mean away.) For the moon or a satellite orbiting Earth, the corresponding terms are perigee and apogee.

An ellipse has two foci (foe-sigh). (Focus is singular, foci is plural.) The sum of the distances of any point on the ellipse to each focus is the same for every point on the ellipse. That is, in the diagram

that follows, $d_1 + d_2$ = a constant. Because of this, one way to draw an ellipse is to place two thumbtacks on a sheet of paper some distance apart (to serve as the foci), tie a loop of string that is somewhat longer than twice the distance between the foci, place a pencil tip inside the string and pull outward to make it taut, and then draw around the thumbtacks while pushing outward to keep the string taut. (If you do the same thing using a single thumbtack, you will instead trace out a circle.)

An ellipse has a major axis and a minor axis (the two bisectors). The semi-major and semi-minor axes refer to half of these.

The Earth's orbit is actually much more circular than depicted in the following diagram.

The mathematical equation for an ellipse is

$$\frac{x^2}{a^2} + \frac{y^2}{b^2} = 1,$$

where a and b are the semi-major and semi-minor axes, respectively (see the following diagram).

(Make sure that you distinguish between the words "ellipse" and "eclipse.")

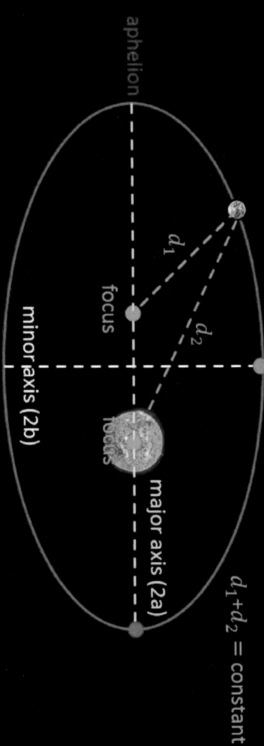

7.2 Kepler's Second Law

According to Kepler's second law, a planet sweeps out equal areas in equal times during its orbit around the sun. Kepler's second law actually holds for all types of two-body orbits in general – including parabolic orbits. It also applies to moons and other satellites orbiting planets, and to comets and other celestial objects orbiting the sun.

In order to sweep out equal areas in equal times along an elliptical orbit, a planet must move faster near perihelion (where it is closest to the sun) and slower near aphelion (where it is furthest from the sun).

Similarly, a comet moves very fast near the sun, and so spends very little time near the sun; it spends most of its time very far from the sun, where it travels very slowly.

The following diagram illustrates that a planet travels faster near the sun and slower far from the sun in order to satisfy Kepler's second law: The planet must cover more distance in the same time near the sun in order to sweep out the same area that it does during the same time interval far from the sun.

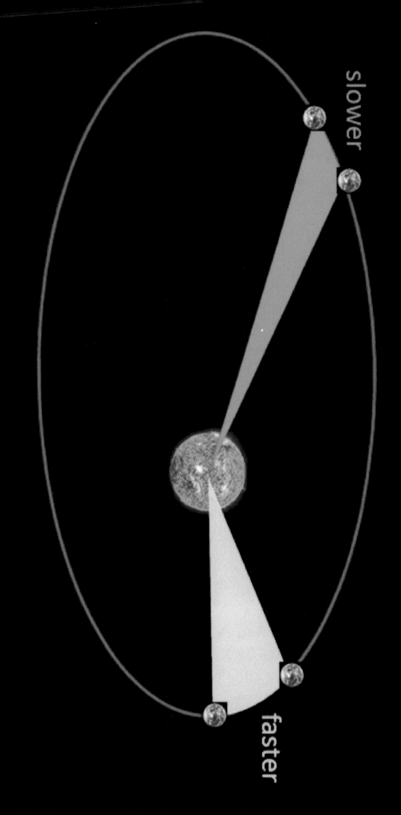

7.3 Kepler's Third Law

According to Kepler's third law, the square of the orbital period (T) of a planet is proportional to the cube of its semi-major axis (a). Recall that the semi-major axis was defined and illustrated in Section 7.1 (the diagram actually labeled the major axis, which is twice the semi-major axis). The orbital period is the time that it takes for the planet to complete one revolution around the sun. As an equation, Kepler's third law can be expressed as

$$T^2 \propto a^3$$

where the symbol $\propto$ expresses proportionality (rather than equality).

Kepler's third law also applies to the moons of a planet. However, you must compare planets that orbit the same star or moons that orbit the same planet when using Kepler's third law. You can't compare the moon's orbit around the Earth to the orbit of Venus around the sun, for example.

Kepler's third law explains why Mercury has the shortest 'year' and why Neptune has the longest 'year,' for example. The greater the semi-major axis of the planet's orbit, the greater its orbital period.

7.4 Newton's Law of Gravitation

According to Newton's law of universal gravitation, any two objects that have mass attract one another gravitationally according to the following equation:

$$F = G\frac{m_1 m_2}{R^2}$$

F is the gravitational force, m_1 and m_2 are the masses, R is the separation between the centers of the masses, and G is the gravitational constant: $G = 6.67 \times 10^{-11}\ \text{N} \cdot \text{m}^2/\text{kg}^2$.

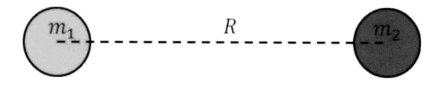

Newton's law of gravity is an inverse-square law. Increasing the separation between two masses decreases the force by a factor of R^2. For example, doubling the separation makes the force four times smaller. Any two objects that have mass attract one another gravitationally. However, in practice, the gravitational force is only significant if one of the masses is astronomical.

Don't confuse the gravitational constant, G, with gravitational acceleration, g. The gravitational constant equals $G = 6.67 \times 10^{-11} \text{ N} \cdot \text{m}^2/\text{kg}^2$ everywhere in the universe. Gravitational acceleration equals $g = 9.81 \text{ m/s}^2$ near Earth's surface, but has other values at other locations in the universe. For example, $g = 1.6 \text{ m/s}^2$ near the moon's surface. These two constants mean much different things, and don't even have the same units.

Gravitational force is only significant if one of the masses is astronomical because G is such a small number (about a trillionth) in SI units.

With a lot of advanced math, Isaac Newton showed an inverse-square law for gravity leads to the following orbits for the two-body system: circle, ellipse, line, parabola, and hyperbola (these are the conic sections that can be obtained by slicing a double-cone with a plane).

Newton's inverse-square law of gravity is a generalization of Kepler's first law. Newton showed that the ellipse is one possible orbit.

Newton also improved upon Kepler's first law: The sun doesn't actually lie at one focus. Rather, the planet-sun center of mass lies at the focus in the restricted two-body problem. The sun is so massive that the difference is very slight.

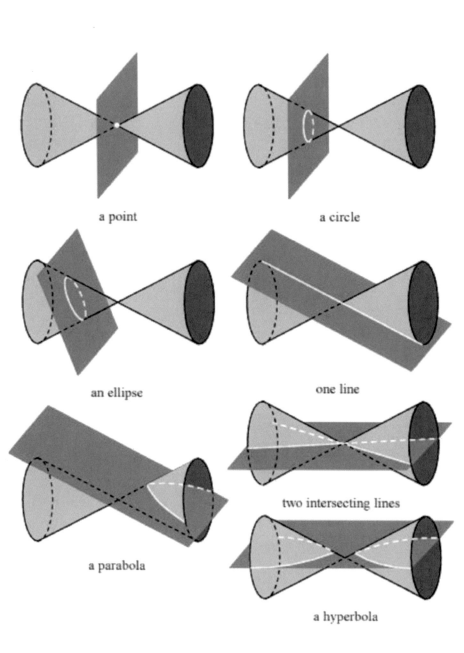

a point

a circle

an ellipse

one line

a parabola

two intersecting lines

a hyperbola

Real planetary orbits are not perfect ellipses because the other planets and moons also have a small effect on the orbits.

A projectile is an object that travels through the air, like a rock that is thrown or a bullet that is shot. A satellite is an object in orbit. The moon is a satellite of the Earth, and the Earth is a satellite of the sun. The Earth also has several man-made satellites.

If a projectile is launched with a high enough speed, it could become a satellite, as illustrated below. A satellite is basically a projectile that's moving fast enough that it doesn't land on the ground. Galileo realized that a satellite is like a projectile, and drew a figure similar to the diagram on the right of the following picture in one of his notebooks.

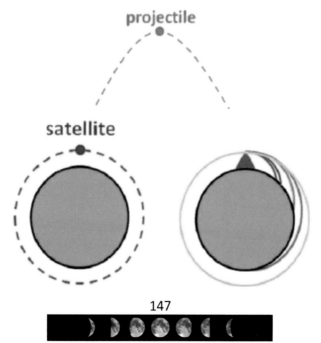

Consider a satellite in a circular orbit. Such a satellite travels with uniform circular motion (i.e. it has constant speed). The gravitational pull of the planet supplies the centripetal force (pulling the satellite inward in order to make it deviate from its inertia – its natural tendency to go off on a tangent, since it wants to travel in a straight line rather than a circle).

All objects inside the satellite have zero apparent weight. They experience weightlessness. Astronauts in such a satellite or space station would feel weightless – they would be floating relative to the satellite, along with any loose items inside the satellite.

The satellite is in free fall, just like an elevator car where the cable has been cut. The difference is that the satellite orbits a planet, and doesn't crash into the ground. The reason is that the satellite has enough tangential speed to complete its orbit (recall the previous comparison between projectiles and satellites).

The Earth attracts the moon gravitationally. Why doesn't the moon crash into the Earth? Because it has enough tangential speed – and inertia (it doesn't want to lose that tangential speed).

7.5 Conservation of Energy

Energy is the ability to do work. There are two basic kinds of energy – potential energy and kinetic energy.

Potential energy (PE) is work that can be done by changing position. When a planet is farther from the sun, it has more gravitational potential energy.

Kinetic energy (KE) is work that can be done by changing speed. When a planet travels faster, it has more kinetic energy.

The total energy of a planet is conserved (i.e. constant): $PE + KE =$ constant.

When a planet is closer to the sun, its potential energy is smaller and its kinetic energy is greater (i.e. it moves faster); when it is farther form the sun, its potential energy is greater and its kinetic energy is smaller (i.e. it moves slower). In both cases, its total energy is constant. This result (that the planet travels fastest at perihelion and slowest at aphelion) is consistent with Kepler's second law and also with the law of conservation of angular momentum.

7.6 Conservation of Angular Momentum

Recall that all objects have inertia – a natural tendency to maintain constant velocity. Rigid objects also have rotational inertia – a natural tendency to maintain constant angular momentum.

The angular momentum of a planet in its orbit around the sun is $L = mvr$, where L is the angular momentum, m is the mass of planet, v is the instantaneous speed of planet, and r is the instantaneous distance from the sun to the planet.

Kepler's second law is actually a special case of conservation of angular momentum. Since the angular momentum of a planet is conserved (i.e. constant), mvr equals the same value at any point in its orbit. When the planet is closer to the sun, r is smaller and v is greater; when it is farther from the sun, r is greater and v is smaller.

Thus, conservation of angular momentum also shows that a planet travels fastest at perihelion and slower at aphelion.

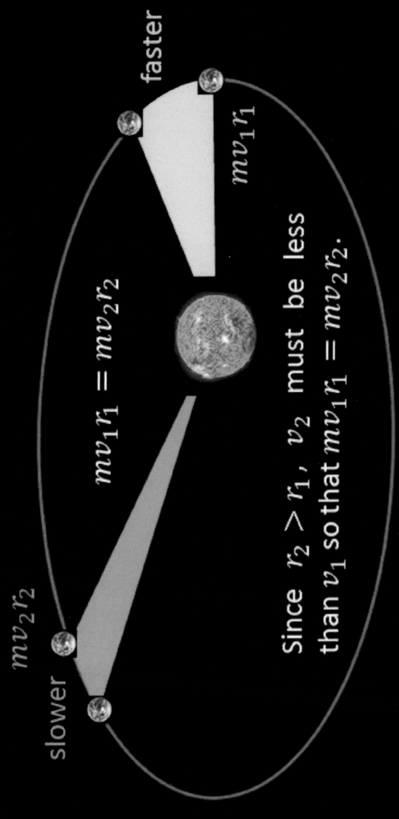

The ice skater's angular speed similarly increases when she brings her arms and leg inward in order to conserve her angular momentum.

7.7 The Earth's Tides

L et us consider one consequence of Newton's law of gravitation.

The tides are the rise and fall of the water level where the ocean meets the land. There are two high tides and two low tides every day.

The sun exerts a greater force on Earth than the moon, yet the moon has a greater influence on tides than the sun because it's much closer to the Earth.

Tides are caused because the moon exerts a greater force on the side of Earth's ocean near the moon than on the side of the ocean far from the moon, causing a squeezing effect on the Earth's oceans.

As Earth rotates, there are high tides when a point is near the moon or far from the moon, and low tides in between. This is illustrated in the following figure (with the effect exaggerated considerably in order to make it easier to see).

Although the moon has a greater effect on Earth's tides than the sun has, the sun's effect does produce a noticeable effect.

High tides are highest and low tides are lowest when the sun, Earth, and moon all form a line.

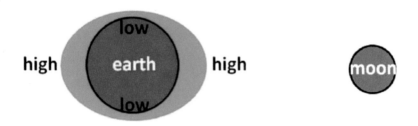

These are called spring tides. (Despite the name, they occur throughout the year, and have nothing to do with the seasons.) These are the maximal tidal effects.

High tides are lowest and low tides are highest when the sun, Earth, and moon form a right angle. These are called neap tides. These are the minimal tidal effects.

There are two spring tides and two neap tides every month, since it takes approximately one month for the moon to orbit the Earth.

Spring tides and neap tides are illustrated in the following figure.

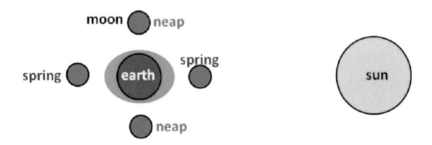

8 Beyond Our Solar System

8.1 The Stars

The nearest star to our solar system is Alpha Centauri. It is about four light-years away from Earth. A light-year is the distance (not a time!) that light travels in one year. Light travels 300,000,000 (three hundred million) meters in one second. The following graph shows the stars in the neighborhood of our sun, with the sun in the center. An arrow indicates which way our sun is headed relative to these other stars.

All of the stars in the northern hemisphere travel in circles around one star, called Polaris, when viewed from Earth. The reason that the stars appear to travel in circles from Earth is that the Earth is spinning on its axis. The stars do not really travel in circles, but are relatively fixed in position relative to our solar system. If you stand at the North Pole and

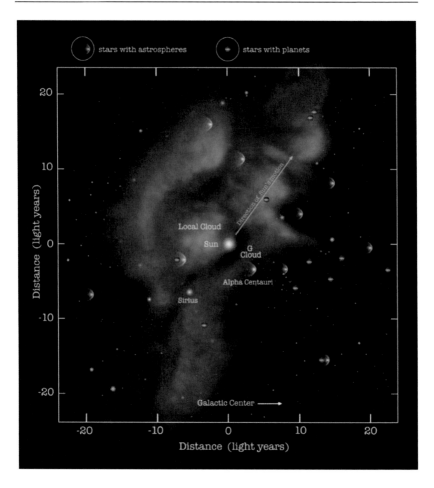

look directly up at night, you will see Polaris. Polaris does not travel in circles, but lies at the center of the apparent circular motions of the other stars, because Polaris is aligned with Earth's axis of rotation. For this reason, Polaris is also called the North Star. Sailors have used Polaris as a useful navigational reference for hundreds of years. The apparent circular motion of the stars due to Earth's rotation is called diurnal motion, and occurs in both the

northern and southern hemispheres. If you setup a camera with a long-time exposure to collect starlight for several hours, the photograph will show this diurnal motion as star trails.

The sun, planets, and moon travel across the sky along roughly the same path because they lie in roughly the same plane. The path that the sun takes across the sky as viewed from Earth is called the ecliptic. Really, the sun is stationary relative to the solar system, and the Earth travels around the sun, but when we view the sun from the Earth, it appears to travel across the sky due to the daily rotation of Earth about its spin axis. The planets and moon are always found not far from the ecliptic. If you want to find planets, pay attention to the points where the sun rises and sets (without looking at the sun – doing so would damage your retinas severely and can cause blindness) and look for planets to be near that same path at night. You can find many online resources to help you find planets and stars by searching for things like, "planets visible tonight," on your favorite internet search engine. You can see that Mercury, Mars, Saturn, the sun, and the moon lie near the ecliptic in the following photo; the ecliptic is that path that approximately connects them.

Stars, on the other hand, appear all over the night sky, not necessarily near the ecliptic. Stars also

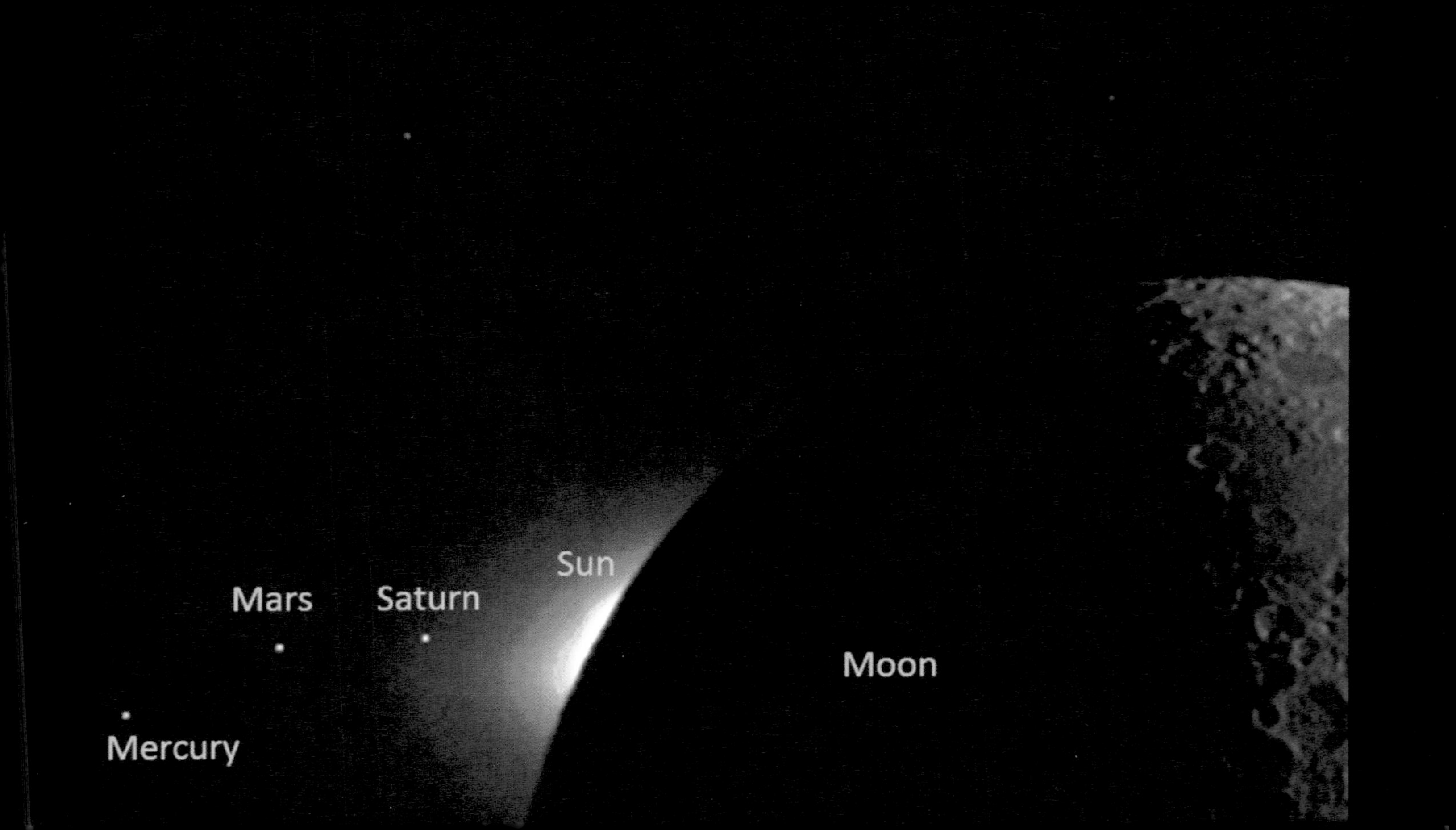
Sun
Mars
Saturn
Mercury
Moon

appear to twinkle, whereas reflected light from the planets does not – though it is not easy to tell which of those tiny lights in the night sky are twinkling and which are not.

Some stars appear brighter than others. This is called apparent brightness. Of course, a star that's actually brighter than another might appear dimmer if it is much further away. The actual brightness of a star is different from the apparent brightness observed from Earth. For the remainder of this book, we will be concerned with the actual brightness of a star, rather than how it appears in Earth's night sky. Astronomers do calculations using the distance of the star from the Earth to determine the star's actual brightness.

If you view stars through a good telescope, you will see that they appear in different colors. Any given star actually emits light in all colors (including infrared, ultraviolet, and other forms of electromagnetic radiation that you don't see with your eye), but you only see one color with your eye that represents a weighted average of the colors that the star emits. A star that appears blue is hotter, while a star that appears red is cooler. This is because the temperature of the star is related to its wavelength through Wien's law, which says that temperature times wavelength is a constant. Higher temperature means shorter wavelength and lower

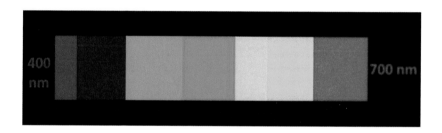

temperature means longer wavelength. Red light has a longer wavelength than blue light, so blue stars are hotter and red stars are cooler. The wavelengths of light decrease from red to violet according to the acronym Roy G. Biv (red orange yellow green blue indigo violet).

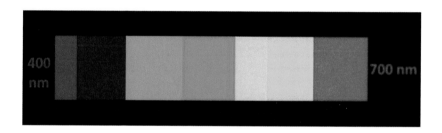

Our sun is yellow, with a surface temperature of 5800 Kelvin. The hottest blue stars have surface temperatures up to about 50,000 Kelvin, while the coolest red stars have surface temperatures as low as 3000 Kelvin.

The spectrum of a star can be seen by pointing a telescope directly toward it and passing this starlight through a prism. The light passing through a prism experiences dispersion – i.e. the different colors that make up the starlight spread out in different directions when passing through a prism, as illustrated on the following page.

Each star's spectrum is unique. It is a signature that astronomers can read to determine the composition of the star.

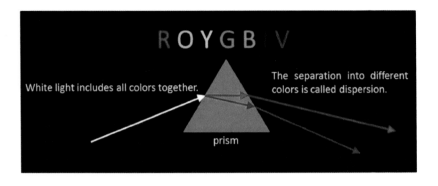

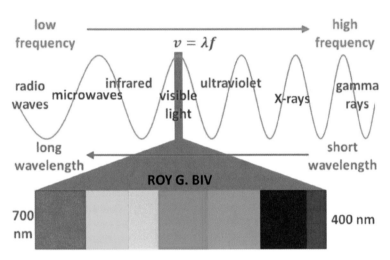

If the star is heading toward or away from Earth, its atomic spectrum will be shifted according to the Doppler effect. When the source of a wave (in this case, the wave is light and the source is the star) is in relative motion with an observer (we observe the starlight from Earth), the frequency is shifted. When the star and Earth are getting closer together, the frequency increases, and when they get further apart, the frequency decreases. Since wave speed = wavelength times frequency, and since

the speed of light is a constant, higher frequency means shorter wavelength and lower frequency means longer wavelength. The following diagram shows that the wavelength is shorter ahead of a moving source and longer behind a moving source (red wavefronts shown on right) compared to a stationary source (blue wavefronts shown on left).

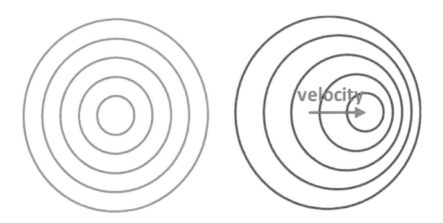

When a star is heading away from Earth, the frequency of the light that it emits is decreased, which means that the wavelength is increased. We call this a red shift since red stars have the longest wavelengths.

When a star is heading toward Earth, the frequency of the light that it emits is increased, which means that the wavelength is decreased. We call this a blue shift since blue stars have the shortest wavelengths.

(Every star emits light of all wavelengths, but when you look at a star with your eye, you just see

one wavelength which represents a weighted average of its spectrum. All of the wavelengths are either red- or blue-shifted, not just one. When we say that red stars have the longest wavelengths, we are speaking just of the average that we see with our eyes. When we say that light is red-shifted, we mean that the entire spectrum experiences a red shift.)

Some solar systems have two stars instead of just one. We call such a system a binary star system. It turns out that about one-third of the solar systems in the Milky Way galaxy (that's ours!) are binary or multiple-star systems.

There are three types of binary systems – visual, eclipsing, and spectroscopic. When we see one star orbit another, we call this a visual binary. Sometimes we don't see two separate stars, but we can infer that light is actually coming from a binary star system. In the case of an eclipsing binary, one star passes in front of or behind the other periodically, which causes periodic variations in the observed brightness. This is called an eclipsing binary. In a spectroscopic binary, we see two sets of spectral lines merged together, and see one set shift relative to the other due to the Doppler effect. As one star orbits the other, it is sometimes red- or blue-shifted compared to the heavier star if it is sometimes going away from or towards the Earth in its orbit.

8.2 The Fate of the Stars

Stars fuse hydrogen nuclei together to form helium nuclei in a reaction called nuclear fusion. It takes four hydrogen nuclei to form one helium nucleus. A hydrogen nucleus has one proton, while a helium nucleus has two protons and two neutrons. The difference in mass between four hydrogen nuclei and one helium nucleus equates to energy according to Einstein's famous equation, $E = mc^2$. Stars are such massive objects that they can fuse four hydrogen nuclei into helium nuclei at a rate of millions of tons each second, and continue to do this for more than a billion years!

Eventually, every star runs out of hydrogen to fuse into helium. When a star runs out of hydrogen, it becomes larger (less dense) and redder (cooler), turning into a giant (often red) or a supergiant. When the core runs out of hydrogen, hydrogen begins to fuse together in a shell around the core.

Why does the star expand into a red giant when it runs out of hydrogen? Ordinarily, there is a balance between the outward radiation pressure (the star is emitting heat and light) and the very strong inward pull of gravity. When the star runs out of hydrogen to burn, it emits radiation at a much lower rate, and gravity shrinks the core of the star

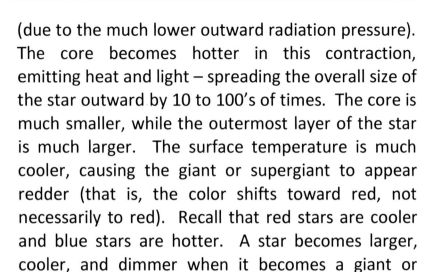

(due to the much lower outward radiation pressure). The core becomes hotter in this contraction, emitting heat and light – spreading the overall size of the star outward by 10 to 100's of times. The core is much smaller, while the outermost layer of the star is much larger. The surface temperature is much cooler, causing the giant or supergiant to appear redder (that is, the color shifts toward red, not necessarily to red). Recall that red stars are cooler and blue stars are hotter. A star becomes larger, cooler, and dimmer when it becomes a giant or supergiant.

If the core becomes hot enough, helium nuclei begin to fuse together to form carbon. If so, eventually the helium runs out. All stars eventually reach a point where they can no longer perform nuclear fusion. There are a few possible fates of a star, depending on the star's mass.

If the star has less than 1.5 times the mass of our sun, it will become a white dwarf. If it has 1.5 to 3 times the mass of our sun, it will become a neutron star. If it has more than 3 times the mass of our sun, it will become a black hole.

A star that has less than 1.5 times our sun's mass will become a white dwarf when nuclear fusion has ceased. This is the ultimate fate of most stars, including our sun. When nuclear fusion ceases, only the core of the star remains behind. Since the core

of a star is very tiny, white dwarfs are tiny stars. A star with the sun's mass will become a white dwarf as small as the Earth. A more massive star (yet still less than 1.5 times our sun's mass) will become an even smaller white dwarf.

The core of a star is the hottest part of the star. For example, our sun's core has a temperature of 15,000,000 (15 million) Kelvin, whereas its surface temperature is 5800 Kelvin. The surface of a red giant is cool, but the core that it leaves behind when it forms a white dwarf is still extremely hot. The reason that a white dwarf still emits enough light to see, and why it looks white, is because it is still the extremely hot core of a star. The white dwarf does cool off in time, and so it grows dimmer and cooler – it is a gradual reduction from the original millions of degrees temperature of the original core.

A white dwarf is extremely dense: It has the mass of a star, but the size of a terrestrial planet (or less). Imagine half the mass of the sun packed into the size of the Earth. Recall that density is mass per unit volume. The white dwarf still has a lot of mass, but much less volume, and so it is much more dense than the original star. To imagine this extreme density, imagine finding a rock the size of a gumdrop that you couldn't even lift! A white dwarf is typically composed of carbon and oxygen (since helium fused together to form carbon and oxygen in the core of the giant that eventually turned into a white dwarf).

Our sun is about 4.5 billion years old. Over the course of the next 5 billion years, our sun will grow about 10% brighter every billion years. In about 5 billion years, it will finally run out of hydrogen to burn and it will turn into a red giant. It will be so large that the sun's new giant size will extend nearly to Earth's orbit. (But you don't have to worry about that for another 5,000,000,000 years.) When nuclear fusion stops, it will turn into a white dwarf. It will take a few trillion years for the super-hot white dwarf to eventually cool down to the very low average temperature of the universe.

A star that has 1.5 to 3 times our sun's mass will become a neutron star. As the name suggests, it is made out of neutrons. They are much more dense than white dwarfs (and recall that a white dwarf was itself very dense). A neutron star has more mass than a white dwarf, but is even smaller. Whereas a white dwarf is about the size of a small terrestrial planet, a neutron star is about the size of a city – yet more massive. To imagine how dense a neutron star is, imagine finding a pebble that weighed as much as a mountain! Gravity would be insanely strong near the surface of a neutron star.

(Not sure what a neutron is? Here is a brief refresher: Protons, neutrons, and electrons make up atoms. Protons and neutrons reside in the nucleus, which is surrounded by electrons. Protons are positively charged, neutrons are neutral, and

electrons are negatively charged. Protons and neutrons are much heavier than electrons.)

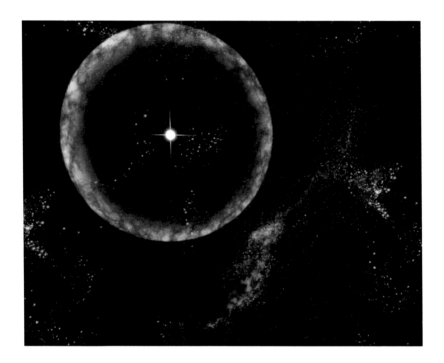

Neutron stars rotate very rapidly. When a slowly spinning mass collapses, its angular speed increases in order to conserve angular momentum (like a spinning ice skater who spins more rapidly when she pulls her arms inward). The concept of angular momentum was discussed in Section 7.6. The neutron star collapses to a much, much smaller radius than the original star, which results in a very large increase in its angular speed. A neutron star can rotate as fast as a hundred rotations per second.

A neutron star emits radiation along the north and south poles of its magnetic axis. If the magnetic axis is not aligned with the axis of rotation, this beam of radiation swirls around like a stellar lighthouse – with a high frequency since the neutron star rotates rapidly. Such a neutron star is called a pulsar. If the beam happens to point toward Earth as it rotates, we see a pulsating signal (we see it like a "stellar lighthouse" rotating with a high frequency). Hence the name pulsar.

If the star has more than 3 times the mass of our sun, it will become a black hole. A black hole is even more dense than a neutron star (which was already extremely dense). When a star has more than 3 times the mass of our sun and nuclear fusion ceases, the star collapses down to the size of a point (called a singularity). Gravity is so strong that not even light can escape it – which is why we call it a 'black' hole. The boundary of a black hole is called the event horizon – beyond the event horizon is the point of no return. The size of a black hole, based on the event horizon, is somewhat smaller than a neutron star (which itself was the size of a city).

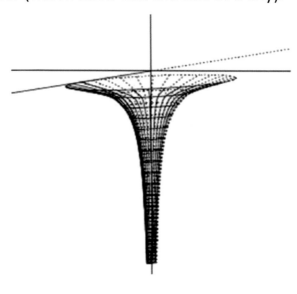

A supernova is the incredibly energetic explosion of a star. Shortly after the explosion, a supernova can appear brighter than a galaxy. During a supernova, a star can release more energy than a star like our sun emits in the course of 10 billion years. Following is the image of a supernova.

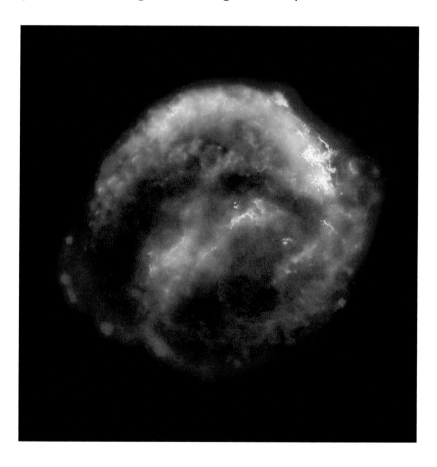

One way for a supernova to occur is for a very massive aging star to experience a sudden gravitational collapse, releasing energy explosively in

the process. A white dwarf that is part of a binary system can also trigger a nova or supernova. In this case, the white dwarf draws mass from the companion star of the binary until its core temperature is high enough to ignite carbon fusion – which becomes a nuclear chain reaction.

Stars are born from a nebula – a large astronomical cloud. The following photo shows a nebula. Gas and dust in the nebula clump together. These clumps collide together, forming larger clumps. Eventually very massive clumps form and the solar nebula collapses gravitationally. Three things happen when a solar nebula collapses. (1) It becomes hotter due to conservation of energy. As the nebula collapses, the gravitational potential energy decreases, so the kinetic energy increases in order to conserve energy (see Section 7.5). Since temperature is a measure of the average kinetic energy of the molecules, this means that the nebula becomes hotter. (2) The solar nebula also spins faster as it collapses in order to conserve angular momentum – like the spinning ice skater who spins faster when she brings her arms inward (see Section 7.6). (3) More clumps collide together, forming yet larger clumps. These collisions cause the solar nebula to flatten into the shape of a disc.

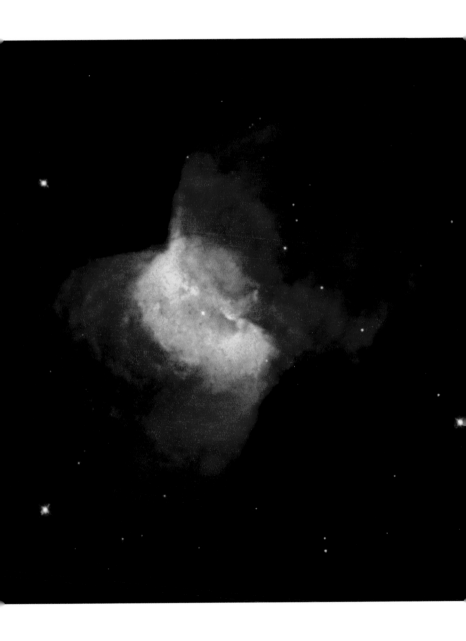

8.3 Galaxies

A solar system is an independent system of planets with one or more stars. A galaxy is an independent system of millions or billions of stars. Our solar system is part of the Milky Way galaxy. It is so named because we see a milky band across the night sky when we look within the plane of our galaxy. The light that makes up this milky band comes from over 100,000,000,000 (100 billion) stars. That's just our galaxy. There are billions of galaxies! The Milky Way measures about 100,000 light-years across (it takes light one year to travel a distance of one light-year). Note that our sun does <u>not</u> lie at the center of the Milky Way galaxy in the following picture, but lies in one of the spiraling arms. The center of the galaxy lies where the spiraling arms merge – find the galactic bar in the photo to find the center of our galaxy.

There are three main types of galaxies – spiral galaxies, elliptical galaxies, and irregular galaxies – which are named for their shape. The Milky Way is a spiral galaxy. We also see some galaxies which are a combination of two types – like partly spiral and partly elliptical, for example. The different shapes of galaxies correspond to galaxies at different stages in their development.

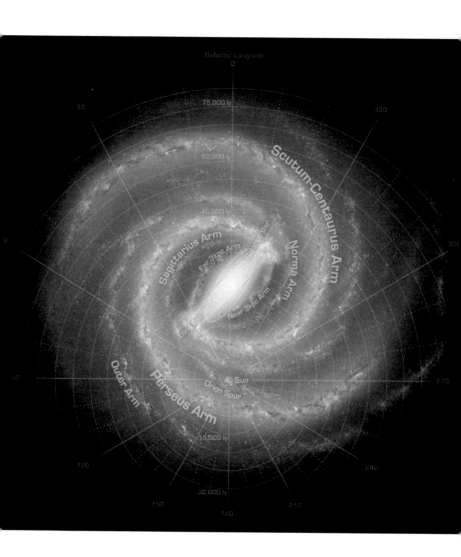

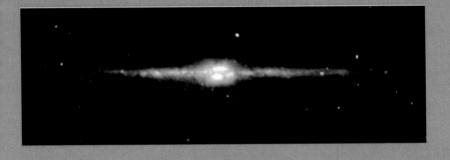

Brief Summary of Important Ideas

- Our solar system is heliocentric, <u>not</u> geocentric. That is, the Earth revolves around the sun; the sun does <u>not</u> revolve around the Earth. We know that the solar system is sun-centered and <u>not</u> Earth-centered because we have observed a full Venus through a telescope, which is only possible in the heliocentric model.
- The order of the planets from the sun is Mercury, Venus, Earth, Mars, Jupiter, Saturn, Uranus, and Neptune. Pluto is now considered to be one of several dwarf planets and one of numerous Kuiper Belt objects instead of a planet.
- The Earth rotates on its (tilted) axis once every 24 hours (one rotation). The Earth orbits the sun every 365 days (one revolution).
- The tilt of Earth's axis causes the seasons.

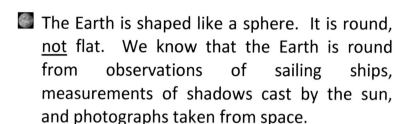
- The Earth is shaped like a sphere. It is round, not flat. We know that the Earth is round from observations of sailing ships, measurements of shadows cast by the sun, and photographs taken from space.
- The moon orbits the Earth every 29.5 days. The phases of the moon are caused by the ever-changing positions of the moon and sun relative to the Earth. They are not caused by shadows of the Earth cast on the moon.
- A solar eclipse occurs when the moon blocks sunlight from reaching the Earth, while a lunar eclipse occurs when the Earth blocks sunlight from reaching the moon. Eclipses are rare because the moon's orbit is tilted compared to the Earth's orbit.
- According to Kepler's first law, the planets orbit the sun along the path of an ellipse, with the sun lying at one focus. According to Kepler's second law, a planet sweeps out equal areas in equal times during its orbit around the sun. According to Kepler's third law, the square of the orbital period of a planet is proportional to the cube of its semi-major axis.

- A planet travels faster at perihelion (closest approach to the sun) and slower at aphelion (furthest point from the sun) according to Kepler's second law. The inner planets complete their revolutions in less time and the outer planets complete their revolutions in more time according to Kepler's third law.

- Alpha Centauri is the nearest star to our solar system. It is about 4 light-years away. A light year is the distance (<u>not</u> time) that light travels in one year. The speed of light is 300,000,000 m/s.

- The name of our galaxy is the Milky Way. The Milky Way is a spiral galaxy.

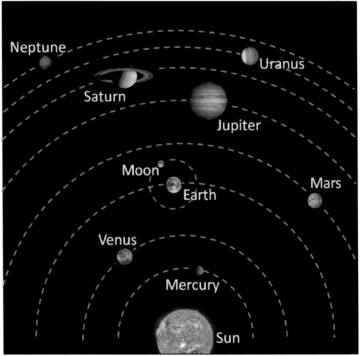

About the Author

C hris McMullen is a physics and astronomy instructor at Northwestern State University of Louisiana. He earned his Ph.D. in physics at Oklahoma State University in phenomenological high-energy physics (particle physics). His doctoral dissertation was on the collider phenomenology of superstring-inspired large extra dimensions, a field in which he has coauthored several papers.

10 9 8 7 6 5 4 3 2 1 0 ...

BLAST

OFF!

Made in the USA
Middletown, DE
30 August 2023

37629652R00104